T0324782

Ecological Design of Smart Home Networks

Related titles

Wearable electronics and photonics
(ISBN 978-1-85573-605-4)

Handbook of advanced dielectric, piezoelectric and ferroelectric materials
(ISBN 978-1-84569-186-8)

Microjoining and nanojoining
(ISBN 978-1-84569-179-0)

Woodhead Publishing Series in Electronic and Optical Materials: Number 70

Ecological Design of Smart Home Networks

Technologies, Social Impact and Sustainability

Edited by

Nobuo Saito and David Menga

AMSTERDAM • BOSTON • CAMBRIDGE • HEIDELBERG
LONDON • NEW YORK • OXFORD • PARIS • SAN DIEGO
SAN FRANCISCO • SINGAPORE • SYDNEY • TOKYO
Woodhead Publishing is an imprint of Elsevier

WP
WOODHEAD
PUBLISHING

Woodhead Publishing is an imprint of Elsevier
80 High Street, Sawston, Cambridge, CB22 3HJ, UK
225 Wyman Street, Waltham, MA 02451, USA
Langford Lane, Kidlington, OX5 1GB, UK

Notice
No responsibility is assumed by the publisher for any injury and/or damage to persons
or property as a matter of products liability, negligence or otherwise, or from any use or
operation of any methods, products, instructions or ideas contained in the material herein.
Because of rapid advances in the medical sciences, in particular, independent verification
of diagnoses and drug dosages should be made.

British Library Cataloguing in Publication Data
A catalogue record for this book is available from the British Library.

Library of Congress Control Number: 2014959680

ISBN 978-1-78242-119-1 (print)
ISBN 978-1-78242-124-5 (online)

For information on all Woodhead Publishing publications
visit our website at http://store.elsevier.com/

Typeset by TNQ Books and Journals
www.tnq.co.in
Printed and bound in the United Kingdom

Working together
to grow libraries in
developing countries

www.elsevier.com • www.bookaid.org

Contents

List of contributors

E.A. Heredia Advanced Technology Lab, Samsung Research America, Mountain View, CA, USA

M. Hirahara Steering Committee of ECHONET Consortium and Toshiba Corporation, Minato-ku, Japan

N. Ishikawa Komazawa University, Setagaya, Tokyo, Japan

M. Isshiki Kanagawa Institute of Technology, Atsugi, Japan

M. Koch devolo AG, Aachen, Germany

T. Minemura Toshiba Corporation, Minato, Japan

T. Murakami Technical Committee of ECHONET Consortium and Panasonic Corporation, Osaka-fu, Japan

S. Owada Sony Computer Science Laboratories, Inc., Shinagawa-ku, Japan

N. Saito Keio University, Minato, Tokyo, Japan

K.J. Turner University of Stirling, Stirling, Scotland, UK

M. Umejima Keio University, Minato, Tokyo, Japan

Woodhead Publishing Series in Electronic and Optical Materials

Part One

Principles and technologies

The concept of an ecological smart home network

N. Saito
Keio University, Minato, Tokyo, Japan

1.1 Introduction

In this age of science and technology, the global economy has developed so much that our lifestyles are now extremely modernized and developed. In some ways, modern society seems to have reached the utmost state of advancement in various areas, including economic development, science and technology pursuit, and the utilization of the given natural environment. However, it is important to consider approaches that may allow human beings to stay longer on the Earth while enjoying fulfilling and peaceful daily lives.

In this chapter, the historical transition of the surroundings of human beings is first described, and some considerations for the problems of this development are discussed. Then, the design approaches to the smart home are listed, along with the problematic components for these designs. The technological components to achieve a smart home design are then discussed, and the difficulties of their implementation are analyzed. Sustainability is necessary for humans to enjoy modern advanced lives because the Earth's resources are limited, particularly with regard to physical materials. Because it is very important for humans to achieve sustainability, related issues are also discussed.

1.2 Historical transitions

1.2.1 Ancient times

The history of humans on Earth started 400,000 years ago, whereas modern human beings started 40,000 years ago. Throughout history, nature supported the lives of human beings. There were a lot of interactions between human beings and the surrounding nature. Our ancestors got food and energy from the environment for the purpose of supporting their lives.

Compared to other animals living on Earth, our ancestors had some level of intelligence. They manufactured tools to make their lives fruitful and convenient. They also made use of language to facilitate useful communication with other humans. It has been stated that the major difference between human beings and other animals is the ability of language use.

Our ancestors started to have families, resulting in their descendants. They then started to organize a number of families into villages or other living spaces. The larger

Ecological Design of Smart Home Networks. http://dx.doi.org/10.1016/B978-1-78242-119-1.00001-1

their organizations were, the stronger they behaved. If there was a struggle between organizations, the stronger one could conquer the weaker one.

Our ancestors started to cultivate the Earth, growing a variety of plants to provide a reliable source of food. They also systematically hunted wild animals for their food or raised useful domesticated animals for various purposes. Our ancestors invented various kinds of tools and devices for food preparation, animal hunting, and plant cultivation and harvesting. Although some animals can make devices using stones and wood branches, humans invented sophisticated tools and machines, even devices that are used to make other tools (i.e., meta tools).

Approximately 10,000 years ago, our ancestors started to form countries and splendid cultures in Egypt and surrounding areas. Ancient Egypt's population became larger and larger, with the culture and civilization developing accordingly. There were powerful kings to govern these countries and their people. By this point, the intelligence of humans had advanced significantly, and many technologies had been developed. The Egyptian technologies included the survey of lands, the observation of sun and stars, building construction (including the pyramids), the management of rivers, making paper, publishing books, and so on.

The so-called Western culture started in Greece and Rome approximately 2500 years ago. The Ancient Greeks were good at philosophy and developed ways of thinking, whereas the Roman people were good at technology. These characteristics were the central points of successive Western culture and society, and they have been influencing the entire course of human history.

There was not much progress in human intelligence and knowledge in the Middle Ages. However, a sudden change occurred in the middle of the eighteenth century in the Western world – the "Industrial Revolution," which greatly impacted the status and behavior of humans. The global population was 800 million in the middle of the eighteenth century, increasing to more than 2 billion by 1945. Another industrial revolution occurred around 1930 with motorization. Human intelligence was then concentrated to invent many new mechanisms – so-called "modern technology." This technology was the base of many new businesses and industries, where great profits were gained.

Figure 1.1 shows the relationship among three important components in history: human beings, nature, and intelligence. The interrelationship among these three components is very vague and shallow, and each of these components has historically

Figure 1.1 Fundamental components of the Earth (acting independently).

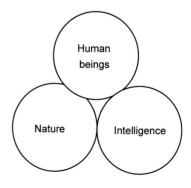

operated rather independently, without considering the mutually dependent relationships and interactions.

1.2.2   Modern times

After the end of World War II, information and communications technology (ICT) started to become practical and even widely used in some areas. Computers essentially are meant to simulate the human brain, although the reality has been much further from an actual human brain's abilities. Nonetheless, ICT was a new and innovative field of technology, as shown in Figure 1.2. Materials science and energy technology were some of the first fields to be developed, even before World War II.

At the beginning, computers were mainly used to calculate numerical data; even so, they were useful for complex mathematical data handling, such as census data. Computers were also used to store and retrieve large amounts of data in business activities, and large-scale business management systems were developed (and prevailed). These computer systems were called *electronic data processors.*

One important advancement in particular brought about the current state of information technology: semiconductor technology development. The transistor was invented by Shockley and colleagues in the 1950s; this element was key in the development of microsized, portable, and customized computer systems. Another critical development was network technology, which led to the creation of the Internet and web systems.

Other technological developments occurred after World War II. Automobile technology advanced, with vehicles becoming much faster and more comfortable. The number of automobiles now manufactured annually exceeds 15 million. The future of the automobile industry is focused on electric vehicles and automatically driven cars.

The energy industry has also advanced, not only because automobiles need to consume much gasoline. For this reason, the consumption of oil increased along with the number of automobiles. In addition, modern houses use electricity, so it was necessary to increase electricity production as well. There are several ways to produce electricity, including fossil fuel power, water power, atomic power, solar power, wind power, and geothermal power. In any case, it is necessary to construct efficient electric power plants.

ICT can support typical business management and organizational management needs as well. By using the appropriate software programs, it is possible to support decision-making based on information given by computers. ICT can solve not only

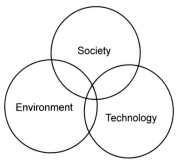

Figure 1.2 Mature components of the Earth (mutual interaction has started).

problems in product manufacturing but also problems in decision-making. Therefore, ICT can support activities related to human intelligence.

ICT can also support financial businesses by using large-scale databases and high-speed data exchanges through well-designed network connections. One can buy and sell stocks using high-speed networks, and one can use banking services through online account access. Online shopping is also prevalent in the daily lives of people of all generations.

Social networking services (SNS), such as Facebook and LinkedIn, now support sophisticated intercommunication among SNS members. These social networks serve as auxiliary communication mechanisms for many members of those services.

As shown in Figure 1.2, technology is very much related to society, with the overlap area between technology and society becoming larger and larger. ICT technology covers mechanisms related to human intelligence, with the above phenomena becoming naturally recognized.

Behaviors in modern society are strongly related to and supported by the economy in many aspects. Modern industries need capital investments; then, in the pursuit of maximum profits, companies need great support from the financial industry. Therefore, the financial industry is a strong driving force in the development of advanced technologies and the growth of the overall economy.

The products supported by modern technologies require much energy and materials, which mainly come from nature. If economic development continues along this path, the Earth's natural resources may be exhausted. Moreover, the natural environment may become contaminated (e.g., air pollution, water pollution). Around 1960, severe environment pollution began to cause deep damage to people's lives. International efforts to reduce pollution around the world are in place, but it has been very difficult to find good solutions to our environmental problem.

1.2.3   The future

The advancement of our technical and economic knowledge is important for all people around the world. It is necessary to change our thinking to realize better solutions for human happiness in the future.

Figure 1.3 shows one of the ways to find a better solution for our future lives. The goal of society is to provide a happy and peaceful life for all citizens, so the family unit

Figure 1.3 Harmonized components of the Earth (full interactions).

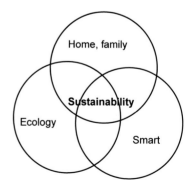

should be respected and well supported. Economic development may be important, but it should be harmonized with natural resource consumption.

"Ecology" is a key concept in efforts to maintain the status of the Earth. If we continue to use automobiles with gasoline engines indefinitely, all of the Earth's fossil fuel reserves will be exhausted. To avoid this disaster, it is necessary to replace current gasoline engines with a new type of engine that uses a fuel other than gasoline. When the fossil fuel is used for heating living spaces, such as buildings or individual houses, ecological considerations would minimize the fuel consumption for heating purposes. Ideally the consumption of fuel would reach as close to zero as possible.

Although advanced technologies will make our daily lives even more convenient, considerations are required to harmonize our lives with the ecology of the Earth. For example, consider the amount of electronic equipment in a modern home: the electric supply for all purposes of energy, the water supply, air conditioning, cooking appliances, hot-water system, communication systems for telephone and Internet, home entertainment system, on-demand entertainment, security system, and so on. As these types of equipment become increasingly affordable, even more people introduce them into their homes to improve their quality of life.

Therefore, it is necessary to consider the direction we are taking. Our intelligence may have produced these modern technologies to make our lives more convenient, but this is not necessarily good for the ecological requirements of the Earth. We need to consider how to use our intelligence for the betterment of both the future of the Earth and the future of human beings. The harmonization of technology and ecology may be called a "smart" consideration – one that is required for human beings to be smarter than in the past.

1.3   Moving toward sustainability

It is important to think about the difference between Figures 1.2 and 1.3. There are three basic components, which have a larger area of interaction in Figure 1.3 than in Figure 1.2. It is important to consider this overlapped area in more detail, noting that the solution for the future (Figure 1.3) is more difficult than for the present (Figure 1.2). It is possible for the overlapped area in Figure 1.3 to be called "sustainability." For example, if a building's air conditioning system has zero energy emission because of the use of sophisticated heating and cooling equipment, this system will not damage the Earth's natural resources, and it is possible to operate this system forever. However, this would not occur in real life. Most modern ways of living consume some amount of natural resources in one way or another. Therefore, this problem requires a more elaborate strategy.

In present and future society, human beings can and will use many technologies to increase economic growth as much as possible, as well as to enhance their lives by maximizing enjoyment, convenience, and comfort. However, it is necessary to use these technologies in a smart or restricted way so that sustainability can be realized. Sustainability means that natural resources are used in such a way that the same usage is assured for future generations. It is possible to reproduce biological resources, but mineral resources cannot be reproduced. Therefore, it is necessary to invent substitutions for mineral resources. Oil and coal cannot be reproduced, so it is necessary to invent artificial oil or gasoline; if this is not convenient or possible, then it is necessary to invent an alternative to gas-powered automobiles.

Sustainability is an important concept and a good goal for our society and technology. It is very smart to maintain the sustainability of our Earth. Because human intelligence is quite different from the intelligence of other living things, as previously discussed, it is our responsibility to protect our world from potential disaster. The smart home is one way to do just that.

1.4 Technology elements

This section presents a short introduction to the technologies that can support the realization of a smart home. It also outlines the necessary environment, with special attention to ecological considerations and sustainability.

1.4.1 Networks, relevant environments, and equipment

A smart home is designed and constructed using high-level ICT and similar technologies. The most important technology is the network, as it is essential to utilize information within the home and among its inhabitants. Information is produced throughout the home, and the rapid exchange of this information without delay is the most important component in an operating smart home. Figure 1.4 shows a common

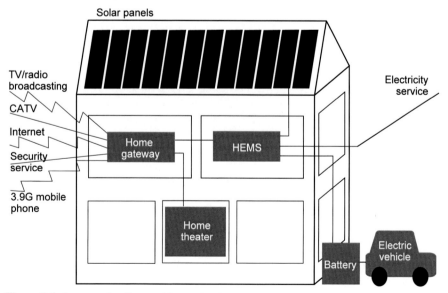

Figure 1.4 An example of a smart home structure.

smart home model, where it is necessary to install an efficient and reliable home network as a base platform.

The necessary technical elements are described briefly in the following sections.

1.4.1.1 Wireless networks and mobile telephones

Wireless networks are now greatly advanced, with most present-day computers being connected through wireless networks. The basic network facility is the local area network (LAN), and its standard is the IEEE802 series (IEEE 802.3). For example, the typical cable-based LAN, Ethernet, is given as IEEE802.3. Wireless LAN is given as IEEE802.11. As wireless communication became more prevalent, another standard organization, the Wi-Fi Alliance (Wi-Fi Alliance), was founded to promote interoperability among different IEEE802.11 communication devices; this wireless communication standard is called Wi-Fi. The name comes from the acronym Hi-Fi, which was used for audio devices.

Starting in the 1990s, mobile telephones began to gain in popularity, and now they can be found all over the world. In 2013, approximately 1.8 billion mobile phones were manufactured in the world, and more than 5 billion people were using mobile phones globally.

The technology of wireless telephone service is quite advanced; phone devices are well designed and sophisticatedly manufactured. So-called smartphones (e.g., iPhones) have become very popular for people of all ages. The smartphone now essentially functions as a handheld computer. It is easy for any user to access the Internet through a browser on a smartphone, so applications can be made easily available through the use of these devices. Mobile phone communication service can be used for the Internet connection service through the use of a tethering service.

In a smart home, there are many applications available, and a mobile phone is one of the interface and operation devices for these applications. Of course, it is necessary to send data in a smart home to the application service provider; therefore, a mobile phone becomes a very important device for smart home inhabitants.

The communication standard for mobile phones is the IEEE 802.11 series, with 3G, 3.9G LTE, and 4G available. For mobile phones, the communication speed is now greater than a gigabyte per second. Therefore, smartphones can be used as handheld computers when connected to the Internet environment.

1.4.1.2 Sensors and their network

It is important to gather various information in the smart home, and sometimes information is gained automatically through several kinds of sensors. Temperature, humidity, motion direction, sounds, and so on are usually measured through sensor devices designed for specific purposes. For example, motion direction can be detected automatically by a strain gage. Sensor device material might be not only metal but also biological material or chemical material. Biosensors may become very useful in the future, and it is expected that superior sensor devices will be developed.

The information collected by sensors should be as accessible as possible. Several simple network mechanisms can be used for data transfer from the sensor devices.

For example, Bluetooth (Bluetooth, 2005) can cover from several meters to several tens of meters for data transfer. Low-energy Bluetooth is now available, and it works with just a small battery. It can be used to connect audio devices and users. ZigBee (ZigBee 802.15.4) and IrDA (infrared data association) are other mechanisms for sensor networks. IrDa can cover 50 cm to 1 m, and it is a rather simple method for a connection between a small device and a computer.

In a smart house, there are uses for mobile phones. A mobile phone can be equipped with a sensor device, such as a strain gage, as it is easy to send these sensor data through a mobile phone connection. It is also possible to send sensor data through Bluetooth or ZigBee to a mobile phone; the data are then sent to a specified service center through the phone's data connection. An integrated circuit card or near-field communication (FeliCa O18092) can be equipped with a mobile phone, working as a node to collect and send integrated circuit card data to the specific Internet node.

1.4.1.3   Internet of things

There is a new area related to the Internet – the "Internet of Things" (IOT) (Holler, Tsiatsis, Mulligan, & Avesand, 2014). It is necessary to extend the Internet connection not only among computers (with human beings) but also among general things (objects). It is estimated that more than 10 billion "things" are connected to the Internet, while 5 billion mobile phones are connected. When computers are connected to the Internet, a receiver computer can respond to a sender computer to facilitate communication between users on the sender and receiver computer. This is based on human habits in the general communication scheme. However, a different situation occurs in the IOT, where the receiver thing itself (not a human being) needs to respond to the sender thing (not a human being). The communication style may be different, and new behaviors and situations occur. The same idea is called M2M (machine-to-machine). The World Wide Web Consortium has proposed a new working group, WOT (Web of Things) (World Wide Web Consortium).

It is, of course, convenient to connect everything to the Internet for managing, utilizing, controlling, neglecting, and so on through the network mechanism. Automobiles, robots, airplanes, trains, ships, and other moving things can be connected to each other through the IOT. Also, houses, buildings, bridges, tunnels, roads, and other constructions can be connected to the Internet. The smart home itself is a target thing to be connected to the Internet, in order to make the management and control of the home easy and efficient.

1.4.1.4   Power line network

It is important to consider ecological principles when a smart home is designed. The main principle is to decrease the cost of construction for a smart home by considering the construction methods, parts, the cost of parts, and so on. When considering the smart home network, it is possible to select an appropriate method and technology to reduce the cost and address any ecological problems. There are choices to be made regarding the network infrastructure, including the power line network structure.

Usually, it is necessary to use electric power in modern homes. Therefore, power lines are installed extensively throughout a home, covering all rooms and facilities to supply electric power to any place in a home. Usually, the base electric power is introduced through electric companies, although recently people have begun to use their own solar cell systems. It may become popular to use electric vehicles to supply electric power to homes through battery systems.

It is possible to use an in-house power line as the LAN infrastructure in the home. It is not necessary to install another line that is dedicated to communication in the home. This can reduce costs and save some material, such as copper. Thus, this approach should contribute to reducing ecological problems on the Earth.

A detailed discussion on power line technology is provided in Chapter 2 in this book. This technology is quite suitable for the ecological design of a smart home network.

1.4.1.5 Femtocell

To provide high-quality wireless communication waves, femtocells (smallcellforum) are used. In each house or office building, femtocells are installed in each room or office. The femtocell itself can be connected to wired Internet cable. It provides superior wireless communication service to users in a smart house or office building.

1.4.1.6 Home network gateway

A smart home needs to use various services provided outside the home, such as Internet communication, telephone and mobile phone communication, television and radio broadcasting, home security service, and emergency communication for natural disasters. Smart home users may be integrating these different kinds of communications one by one, which is time consuming and expensive. It is better to concentrate these communication connections into an integrated gateway. This kind of communication hub is called a "home gateway" (HGW) (Home Gateway Initiative). Currently, HGWs are not optimally designed, but in the near future, some level of standardization will be realized.

1.4.2 Software: platform and user interface

It is very important to use sophisticated system software with the above-mentioned hardware components so that a smart home's ecological design strategy can be well implemented.

1.4.2.1 Operating system

Several systems work as basic components for constructing a smart home and its networks. With these well-designed and popular operating systems (OSs), it is possible to select an appropriate OS for each of the target systems. Linux, Android, and iOS are some of the typical OSs.

1.4.2.2 Platform for a smart home network

A smart home network is constructed by combining and connecting several components, such as baseline networks, sensor networks, HGWs, and special-purpose networks. It is very important to integrate these components into a smart home network platform so that communication among components can occur seamlessly.

1.4.2.3   User interface, user interaction, and operation

Users in a smart home are usually family members, with a variety of age and experience. It is important to make these systems as user-friendly as possible; therefore, a well-designed user interface is essential. The keyboard, push buttons, voice interface, and sound interface are all very important for ease of operation.

1.4.2.4   Security and privacy management

A smart home network supports all family members, so there is a lot of private information from these family members going around the network. Therefore, it is very important to guard against security invasions and ensure each family member's privacy.

1.4.2.5   Standardization for the home network

Smart homes will become more and more popular in the future, with the demand for well-designed smart home networks increasing. Therefore, it is very important to establish standards for smart home networks. Using these standards, it will be rather easy to build a sophisticated smart home and its network. An example of such standardization is given in Chapter 3.

1.4.3   Energy management in the smart home

1.4.3.1   Home energy management system

From an ecological point of view, it is very important to design a smart home that can reduce daily energy use without much effort. The home energy management system (HEMS) is an important concept in this regard. HEMS includes the automated control of electricity usage through a HEMS meter, for example.

It is possible to apply this concept to an apartment building or a group of several houses within a neighborhood. In this case, a *smart grid* (Smart Grid) can be used, which is a new technology to save electricity for families in a wide area.

The standards for home appliance energy management are given in Chapter 6.

1.4.3.2   Electric vehicle

Current automobile technology aims to make the electric car more useful and accessible, so that anyone can use an electric car anywhere. The electric car is the current goal in a compromise between the automobile society and natural environmental ecology. Automobile technologies related to the electric car will likely become greatly advanced in the near future.

The electric car is equipped with a fuel battery, and it is possible to send electricity to a smart home when the electric car is parked in the garage. A solar battery can also be used for this purpose. It is possible to optimize the smart home operation and electric car driving.

1.4.3.3 Remote control of home equipment

In a smart home, appliances such as air conditioners can be switched on and off through mobile devices from outside the home. ECONET Lite, the original Japanese home network standard (see Chapter 6), has defined a remote control scheme for home appliances. Several home appliance manufacturers provide ECONET Lite-based air conditioners, washing machines, and lighting systems with remote control switches, which can support energy saving.

1.4.4 Integrated applications

A smart home network can include a variety of applications for the home's inhabitants, which may include rather complex and integrated applications. This section describes some examples of such applications (Wu & Saito, 2013).

1.4.4.1 Content management in a home theater

A smart home must be equipped with high-quality video screens and high-fidelity sound devices. Family members can use this equipment as a high-quality home theater. It is possible to manage multimedia content in an elaborate way so that one can take in media content from outside the home through the Internet, mobile devices, or HGWs. It is necessary to install high-speed networks to smoothly manage this content. A detailed description is given in Chapter 7.

1.4.4.2 Human activity recognition, personal behavior monitoring, and management

In the coming years, the elderly population will increase around the world. It is necessary to care for older people so that they can stay happy and be safe in their daily lives. ICT and related technologies can be used for the care of elderly people.

In a smart home, rooms can be equipped with video cameras, making it possible to monitor the movement of family members within the house. It is also possible to analyze the video image using pattern recognition technology; then, when a dangerous situation occurs in a house, it is possible to send an alarm signal to a watch station. High-quality pattern recognition technology, high-quality image analysis technology, high-quality video cameras, and a reliable security network can work together to take care of elderly people in a smart house, as shown in Figure 1.5.

1.4.4.3 Health care and medical care

It is also possible to use ICT to monitor an individual's health status. The development of health-related equipment, such as weight meters (called health meters), blood pressure meters, and cardiographs, has been very promising. It is possible to send these measured data to a health service station automatically through the sensor network and middleware infrastructure (see Chapter 3).

It is also possible to co-ordinate food delivery service with a healthy food menu based on these daily health data. A smart house with a high-quality smart network can

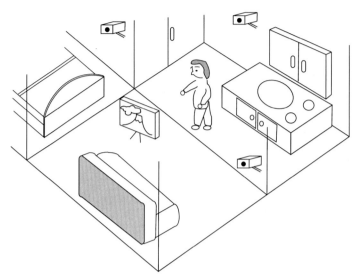

Figure 1.5 Monitoring of an elderly resident in a smart home.

be very important in helping people to live a healthy life, and it would also contribute to reducing the cost of medical care.

1.5   Design considerations

When conceptualizing a smart house or a home network, it is necessary to define the specifications for the components of the project. In this decision-making, one may select a specification if it satisfies the given requirements, then consider several other conditions and objects. The following sections highlight the requirements to be taken into consideration in the design process.

1.5.1   Functional requirements

First, a house should be designed to provide its inhabitants with comfortable and convenient lives. These are basic functional requirements, and some users might want to request additional functionalities such as heating and air conditioning, a home theater system, a physical training facility, and so on.

A home network is also designed to satisfy the users' requirements for network demand and performance. The functions provided by the HGW are key for better home network installation.

1.5.2   Cost and schedule of construction

It is important to consider the cost when constructing a smart house. The introduction of new technologies may cause an increase in cost. A compromise between the budgeted cost and desired technologies may be required. In addition, the construction schedule is also important for users.

1.5.3   Ease of use and safety considerations

For the users of a smart house, the essential problem is the functional requirements, which should be completely installed and implemented. It is important for inhabitants to be able to use these functions without difficulties. It is very important for users that the smart house can be easily maintained and managed.

In addition, a smart house should assure a safe daily life, with some support for safety from natural disasters or accidents. An anti-earthquake design is necessary, especially for areas with a high frequency of earthquakes. A fire prevention system is also very important.

1.5.4   Aesthetics and enjoyment

People tend to like beautiful or pretty objects. Therefore, they expect their houses to be well designed, with an artistic viewpoint.

1.5.5   Ecological considerations and sustainability

There is another way of thinking to be considered: ecological requirements and sustainability. For example, energy consumption and supply are very important considerations in design decision-making. Energy management through HEMS can be a difficult point. Sometimes, the requirements for convenience and ecological restrictions may conflict with each other.

The materials used to build the house should also be considered deeply. Wood materials are easy to use, but they may be restricted by volume when they are natural materials. Steel, plastic, and concrete materials also come from natural resources on the Earth. Material sciences may bring about new ideal materials for better ecological solutions in the future.

Ecological considerations may increase construction costs, whereas users may prefer and even select easy and low-cost materials. However, ecological design does not have to work against the desired technology. If the cost of ecological implementation increases, the technology in question may not be well developed. We need to make an effort to develop technologies to support ecological design and implementation so that our society can enjoy the sustainability of our Earth.

1.6   Future trends

This chapter serves as an introduction to the ecological design of a smart home network, summarizing the historical development of our society. Modern society enjoys the advanced development of technologies, but it is necessary to consider the future of our environment. The key point here is the sustainability of our planet.

It is possible to make the living environment – the home – smart and ecologically satisfying, by using high-end ICT or a smart home network. A smart home leads to a smart building, which can further lead to a smart city and even a smart country in the future.

References

Bluetooth (IEEE 802.15.1). (June 2005). *IEEE standard for information technology—Telecommunications and information exchange between systems—Local and metropolitan area networks—Specific requirements Part 15.1: Wireless medium access control (MAC) and physical layer (PHY) specifications for wireless personal area networks (WPANs).*

FeliCa (ISO18092) ISO/IEC 18092. 2004 information technology – Telecommunications and information exchange between systems – Near field communication – Interface and protocol (NFCIP-1), This standard has been revised by: ISO/IEC 18092:2013.

HGI (Home Gateway Initiative). http://www.homegatewayinitiative.org/.

Holler, J., Tsiatsis, V., Mulligan, C., & Avesand, S. (2014). *From machine-to-machine to the internet of things: Introduction to a new age of intelligence.* Academic Press.

IEEE 802.3 IEEE P802.3.1 Revision to IEEE STD 802.3.1-2011 (IEEE 802.3.1a) Ethernet MIBs.

Smallcellforum. http://www.smallcellforum.org/.

Smart Grid. http://energy.gov/oe/services/technology-development/smart-grid.

Wi-Fi Alliance. http://www.wi-fi.org.

World Wide Web Consortium. http://www.w3.org.

Wu, Z. L., & Saito, N. (Eds.). (November 2013). Special Issue: The Smart Home. *Proceedings of the IEEE, 101*(11).

ZigBee (IEEE 802.15.4) 802.15.4-2011-IEEE Standard for Local and metropolitan area networks–Part 15.4: Low-rate wireless Personal area networks (LR-WPANs).

Power line communications and hybrid systems for home networks

M. Koch
devolo AG, Aachen, Germany

2.1 Introduction

The term *power line communications (PLC)* generally refers to the transmission of data signals via the existing low-, medium-, and high-voltage (LV, MV, HV) electricity distribution infrastructure. This chapter focuses on PLC technologies for home networks. Here, the existing mains grid in the home is used for the transmission of data signals. Each mains socket can become a data port for sending and receiving data signals by simply plugging a PLC modem into it. Typical applications are distribution of digital subscriber line (DSL) signals, television (TV) via Internet (Internet Protocol Television (IPTV)), for audio streams, connecting Webcams, or home/building automation.

The idea of PLC is more than 100 years old. Loubery filed a patent application on this idea in 1899. This first PLC patent was granted in 1901. Since then, PLC technologies have developed from analog to digital, from transmitting only a few bit/s into broadband transmission of up to 1 Gbit/s, from high radiation toward radiation compatible with the electromagnetic compatibility (EMC) environment, and from unprotected transmission toward encrypted transmission on the same security level as mobile network/wireless standards.

Regarding the current trend toward smartphones and tablet PCs, more and more applications are based on wireless Wi-Fi connections. However, because construction and building characteristics vary, Wi-Fi signals might not pass easily through walls or floors in some buildings. A good solution is to transmit the data signal via PLC throughout the building and use a PLC to Wi-Fi converter in the room where the Wi-Fi access point should be made available (see Figure 2.1). This is a simple example for a hybrid system combining PLC and wireless technology.

2.2 Power line technologies

This section provides an overview of the existing PLC technologies. Figure 2.2 shows the three frequency ranges used for PLC data transmission. The lowest frequency range (3–148.5 kHz) is defined by the European standardization platform CENELEC in EN50065. Narrowband PLC technologies operating in this frequency range will be discussed in Section 2.2.1. The frequency range 150–500 kHz is used especially in the United States and therefore is sometimes called

Ecological Design of Smart Home Networks. http://dx.doi.org/10.1016/B978-1-78242-119-1.00002-3

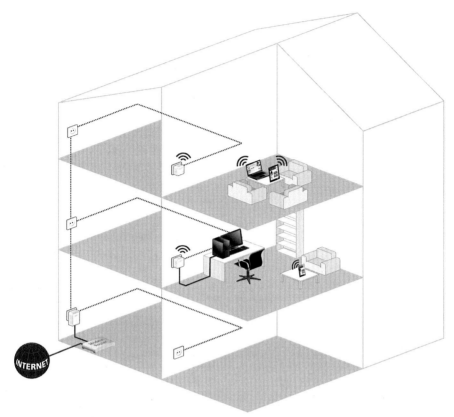

Figure 2.1 Hybrid home network combining power line communications and Wi-Fi. devolo AG.

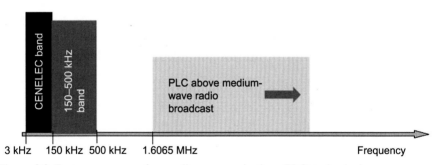

Figure 2.2 Frequency ranges of power line communications (PLC) technologies.

"FCC-band" (Federal Communications Commission is the US regulator) in the literature, which is actually incorrect, as 150–500 kHz is also used in European countries on MV and HV, and might be used on LV in the near future for PLC transmission. It will be discussed in Section 2.2.2. The spectrum 500 kHz–1.6 MHz is not used for PLC transmission because it is used by medium-wave radio broadcasters. PLC that

Table 2.1 **Advantages/disadvantages of the different frequency ranges used for power line communications transmission**

Frequency range	Physical data rate	Distance
3–148.5 kHz	1–80 kbit/s	Up to 2 km
150–500 kHz	100–500 kbit/s	Up to 2 km
1.6065–30 MHz	14–200 Mbit/s	100–200 m
30–300 MHz	Up to 1 Gbit/s	1–10 m

transmits in the third band starting from 1.6065 MHz is often called *broadband PLC* in the literature, which is again not quite correct as some (older) PLC technologies operating here show much lower data rates than others in the 150–500 kHz range. The highest frequency band will be discussed in Section 2.2.3.

Table 2.1 provides, indicatively, an overview of physical data rates and distances. The available data rate on the application layer is much smaller. The stated absolute values should only serve as examples and are permanent matter for improvement of the specific implementation. As one can see, it is always a trade-off between achievable data rate versus distance. When choosing a PLC technology for home networks, solutions above 1.6 MHz are generally favored, as the distances within homes are within the achievable range. However, for building automation of multidwelling units (MDU) or semipublic buildings, a lower frequency range might be necessary to cope with the distances.

Because a three-phase power grid is used in many homes, it is frequently asked whether there is a communication connection possible when the sending PLC modem is plugged into a socket with a different phase than the receiving one. Usually, all wires with different phases run for a couple of meters in parallel in the proximity of the power meter and the fuses. The distances of this parallel wiring are usually enough for cross-talk of the PLC signal from one phase to the other. Therefore, there is no problem with sending and receiving PLC on a different phase – in particular with PLC in higher frequencies. However, if a problem should be identified, a phase coupler for band-passing the used PLC frequency needs to be mounted on the Deutsches Institut für Normung (DIN) rail in the meter/fuse rack.

2.2.1 Narrowband technologies

In Europe, the European Harmonized CENELEC Standard EN50065 provides PLC signal level limits and specifies the use of the frequency spectrum from 3 to 148.5 kHz for different PLC applications. For use, EN50065 defines five different bands, from which only the minority of spectra are allowed for private use in home networks: B, C, and D bands.

- 3–9 kHz: For utility use only.
- 9–95 kHz: For utility use only. Called "CENELEC A band."
- 95–125 kHz: For private use. Called "CENELEC B band."

- 125–140 kHz: For private use. Applied PLC technologies must support a carrier sense multiple access protocol with the beacon carrier at 132 kHz for sharing this spectrum between different users and enabling coexistence of different PLC implementations. Called "CENELEC C band."
- 140–148.5 kHz: For private use. Called "CENELEC D band."

CENELEC EN50065 specifies neither the physical (PHY) layer (layer 1 of the open systems interconnection (OSI) layer model) nor the link layer (layer 2) of the PLC system. Therefore, PLC implementations following EN50065 are either proprietary or follow a "de facto" standard by an industrial association. Interoperability between different PLC technologies cannot be guaranteed. The most widely used PLC technologies are the X-10 PLC Command and Control System, the CEBus standard (also called ANSI/EIA-600; see G. Evans), and the LonWorks PLC specification (Echelon). These PLC implementations are typically based on single-carrier systems using low-complexity modulation schemes such as phase shift keying (PSK) or frequency shift keying. The data rates that are achieved range from a few bit/s up to about 1 kbit/s. More modern technologies based on multicarrier systems will be discussed in Section 2.2.2.

The maximum achievable data performance in the B, C, and D bands is very low due to the size of the bandwidth, the limitation of the PLC signal level, and the high noise rate. The maximum theoretical channel capacity C in bit/s can be estimated by using Shannon's theorem.

$$C = \int_{f_0}^{f_1} lb \left(1 + \frac{S}{N}\right) df \tag{2.1}$$

To become familiar with this theorem, for simplicity it is assumed that the signal-to-noise ratio (SNR) is 5 dB linear over the CENELEC C band of 125–140 kHz. So, in this example the maximum channel capacity, in theory, would be

$$C = 15 \text{ kHz} \times lb \left(1 + 10^{0.5}\right) = 31 \text{ kbit/s} \tag{2.2}$$

The PLC implementations as mentioned above are far from meeting this theoretical performance. Newer approaches for achieving this performance are based on multicarrier systems and will be described in the subsequent chapter.

2.2.2   Multicarrier technologies below 500 kHz

Whereas PLC technologies in the higher-frequency spectrum above 1.6 MHz were based on multicarrier approaches since about 2000, these have been adopted to the lower-frequency spectrum only starting since about 2006. The PLC technologies PoweRline Intelligent Metering Evolution (PRIME) and G3-PLC follow this approach, and have the highest deployment rate at the moment (2013). Both technologies use orthogonal frequency division multiplexing (OFDM), although the parameters are different with each technology.

OFDM is not only used for PLC technologies, but for many years has been used in xDSL, Wi-Fi (IEEE 802.11), digital audio broadcasting, digital video broadcasting, etc. Such OFDM-based implementations need only a relatively low minimum SNR and are robust against phase distortion and impulsive noise. OFDM is a multicarrier technology that divides the available spectrum in to several subcarriers, optimizing the modulation in each one independently. Subcarrier modulation could range, for example, between binary phase-shift keying (BPSK) on a very poor transmission channel up to 4096 quadrature amplitude modulation (QAM) or even higher on a very good transmission channel, depending on the technological implementation. That means each subcarrier does its own channel quality estimation and optimizes the modulation accordingly. Subcarriers are partially overlapping, maximizing transmission capacity. One main parameter describing an OFDM implementation is the bandwidth of a single subcarrier Δf. Another parameter is the number of subcarriers N. With the overlapping of the subcarriers always in maximum one subcarrier to zero passing of the neighborhood subcarrier, the total bandwidth B is calculated to

$$B = (N+1) \times \Delta f \qquad (2.3)$$

For $N \gg 1$ and the symbol rate r_s, the spectral efficiency of an OFDM system can be described by

$$B \approx N \times r_s \qquad (2.4)$$

In comparison, a single-carrier system shows the spectral efficiency of $B > 2 \times N \times r_s$. That means that an OFDM system shows about double efficiency.

PRIME was originally an EC-funded project, launched by Iberdrola and started in 2007. PRIME defines the PHY, the media access control (MAC), and the convergence layer. Therefore, implementations based on the PRIME specification can interoperate. ITU has adopted the PRIME specification for ITU Recommendation ITU-T G.9904. PRIME uses the subspectrum of the CENELEC A band 42–89 kHz with $N=96$ subcarriers and $\Delta f=488.28125$ Hz. It supports differential binary phase-shift keying, quaternary, and 8-ary (D8PSK) modulation. If the transmission channel is the best, D8PSK can be used on all subcarriers, which results in a maximum achievable data rate of 128.6 kilobits per second (kbps) without forward error correction coding and 64.3 kbps with convolutional coding (see PRIME Technology White Paper). In real installation, the achieved data rates are much lower. Therefore, the PRIME Alliance plans to extend its specification up to 500 kHz, but this extension was not available as of August 2013. As the spectrum above CENELEC A is not protected for utility use only, the technology with its extension will also become interesting for private applications such as home or building automation. At the moment, PRIME is only used by utilities, mainly for automatic meter reading (AMR).

G3-PLC is also OFDM based and was originally developed by Maxim to meet the requirements of Electricité Réseau Distribution France (ERDF). G3-PLC technology has been available on the market since 2010. The specification was published by the G3-PLC Alliance and adopted by ITU as Recommendation ITU-T G.9903. It intends to ensure interoperability between different G3-PLC implementations, and addresses similar aspects as PRIME. In contrast to PRIME, the G3-PLC specification defines the technology for using the entire frequency spectrum up to 500 kHz. Therefore, in addition to use in utilities, G3-PLC can already be used for private application such as home and/or building control. In the CENELEC A band, G3-PLC uses the spectrum 36–91 kHz with 36 subcarriers and $\Delta f = 1.5625$ kHz. Above the CENELEC A band, G3-PLC uses the spectrum 145–478 kHz with 72 subcarriers and $\Delta f = 4.6875$ kHz. D8PSK is the highest available modulation. If the transmission channel is the best, D8PSK can be used on all subcarriers, which results in a maximum achievable (physical) data rate of 46 kbps in CENELEC A and 234 kbps in the higher spectrum (150–500 kHz). In real installations, 20–60 kbps has been measured in 150–500 kHz (for test results, see Koch, 2012). These achieved data rates are on the application layer; PHY layer data rates are nominally much higher.

In March 2010, IEEE established Project 1901.2 "Standard for Low Frequency (less than 500 kHz) Narrow Band Power Line Communications for Smart Grid Applications." IEEE P1901.2 also uses the OFDM approach with elements from both PRIME and G3-PLC. The standard was released before the end of 2013. First implementations of IEEE P1901.2 in chips are already available, also as a flexible implementation that can be configured by firmware to support PRIME, G3-PLC, or IEEE 1901.2.

2.2.3   PLC technologies above 1.6 MHz

The starting frequency for PLC transmission of this category of PLC technologies shown in Figure 2.2 is chosen in reference to CENELEC EN50561-1; however, technical implementations might vary. The ending frequency for PLC transmission was originally limited to 30 MHz by standardization and regulatory consideration, but has been extended since about 2010 up to 400 MHz. However, most technical implementations remain below the spectrum of ultrashort-wave broadcast radio services of 87.5 MHz.

Most home networks based on PLC in Europe and North America use the technology specified by the HomePlug Powerline Alliance. The HomePlug Powerline Alliance published its first PLC specification, "HomePlug 1.0," in 2001. Products based on this specification were launched since 2003 and achieved with a maximum physical data rate of 14 Mbit/s, a broad market acceptance at this time. The home network application was mainly the distribution of Internet signals such as received via DSL all over the home.

In 2005, HomePlug published "HomePlug AV" (AV = audio and video) with an increased physical data rate of 200 Mbit/s, which allows the additional distribution of several video streams all over the home in parallel. The HomePlug broadband PLC approach was always based on OFDM.

In the following, the HomePlug evolution was in parallel with the standardization work done in IEEE 1901. In 2005, IEEE started Project 1901 "Standard for Broadband over Power Line Networks: Medium Access Control and Physical Layer Specifications" and approved/published the standard at the end of 2010. The IEEE 1901 standard offers two different PHY layer options with their related PHY-dependent MAC layer. The two PHY options are the HomePlug evolution HomePlug AV2 (first published in 2009) based on OFDM and a wavelet-based approach, mainly developed by Panasonic. Both options can be combined in one PLC device. Both options can achieve a maximum physical data rate of 500 Mbit/s. The European market basically accepts the OFDM option (HomePlug AV2), whereas the wavelet option targets the Asian (Japanese) market.

PLC uses the power grid as a shared medium. In the MDU scenario, the power grid might be used by several people for PLC applications and, because of short distances between different flats, the PLC signals from one flat might be detectable in another. To avoid interference and eventually mutual blocking between different PLC technologies when used in such a scenario, the IEEE 1901 standard specifies an Inter-System Protocol (ISP) to ensure coexistence. This ISP has been developed in liaison with ITU and adopted into the Recommendation ITU-T G.9972. It ensures, in particular, the coexistence between the IEEE 1901 options and the PLC specified by ITU in Recommendation ITU-T G.9960 (also known as ITU G.hn).

HomePlug is permanently trying to improve its technical specifications. The latest published one is HomePlug AV version 2.0, March 22, 2012. It is backward compatible with HomePlug AV and IEEE 1901. HomePlug AV2 uses the frequency spectrum from 1.80 to 86.13 MHz on 3455 subcarriers with the modulation range from BPSK up to 4096-QAM. The achievable data rates are higher than with HomePlug AV, but the main technical achievement is better coverage of all sockets in the home with a higher average data rate (the maximum physical data rate could only be achieved previously with some sockets). The main reason for the better coverage is the introduction of the multiple-input multiple-output (MIMO) concept. The MIMO principle has been used for some time in wireless technologies (e.g., Wi-Fi according to IEEE 802.11n), where more than one sending antenna is connected to more than one receiving antenna. This principle has been adapted to PLC by making use of the protective earth (PE) wire, which is usually available in the home at every socket. MIMO PLC devices transmit on any two wire pairs within three-wire configurations, whereas standard single-input single-output (SISO) PLC always transmits on the line-neutral pair. Whenever the PE wire is not available, HomePlug AV2 automatically switches to standard SISO operation.

Another OFDM-based, PLC technology operating above 1.6 MHz that has been specified and published first in 2010 by HomePlug is called GreenPhy. GreenPhy's design was optimized for the principal applications of monitoring and controlling devices via low-speed and low-cost PLC. In home networks, GreenPhy is interesting in particular for home and building automation. GreenPhy is interoperable with Home-Plug AV, HomePlug AV2, and IEEE 1901. Therefore, applications can be connected via GreenPhy to an existing home network based on these technologies, for example, multiple video streaming. GreenPhy has been designed for low bill of material and low power consumption. These targets are achieved by a reduction of the modulation scheme to quadrature PSK for the best channel (HomePlug AV2 can modulate up to 4096-QAM), so that less powerful CPUs for each carrier can be chosen; the result is lower costs and lower power consumption. The HomePlug Alliance states that Green-Phy offers a minimum effective data rate of 1 Mbit/s, with a peak physical data rate of 3.8 Mbit/s and a maximum peak physical data rate of 10 Mbit/s. In its last version 1.1, published April 11, 2012, GreenPhy uses 1155 carriers, operating in the frequency range 2–30 MHz.

Another interesting application for GreenPhy is as communication channel between an eVehicle (electric vehicle) and electric vehicle supply equipment (EVSE, e.g., battery charger). The standardization for this communication interface is handled in a joint ISO/IEC working group. This ISO/IEC working group adopted GreenPhy into its draft standard 15118-3, which is expected to be approved and released as a standard at the end of 2014, respectively, beginning 2015. Charging a battery at someone's own premises might have the effect of adding this communication as part of the home network. In addition to the pure charging, monitoring the charging from a smart phone attached to the home network via Wi-Fi, for example, might become another use of a hybrid PLC/Wi-Fi home network.

2.3   Hybrid wireless/power line systems

This section describes the benefit, concepts, and application of using hybrid networks, in particular for, but not restricted to, PLC/wireless, for home networks.

2.3.1   Concept of PLC as backbone and available technologies

The concept of using PLC as backbone and connecting it to wireless technologies is motivated by different scenarios and combinations of them. For example:

1. The wireless technology is needed for connecting the terminal, whereas it has difficulties transmitting thorugh floors or walls. Refer to Section 2.1 Introduction and Figure 2.1. For implementation in the home network, there are many broadband PLC-to-Wi-Fi 802.11 bridges available on the market. Usually, such a bridge is simply plugged into the mains socket.
2. The wireless technology is needed for connecting the gateway into the house, but the terminal is connected to PLC. A long-term evolution (LTE = fourth-generation mobile network standard) gateway with outdoor antenna and the distribution of the Internet signal inside the house is such an example. Technical implementations would bridge the LTE signal to broadband PLC, and have been announced to be available soon or are already available.
3. A meter (e.g., gas meter, water meter) has no connection to the power grid, but the AMR should be processed through the same gateway as a power meter with PLC connection – either between private household and utility or inside the home to a central display (e.g., on a PC or smartphone). This meter usually offers a wireless interface. Depending on the region and/or the decision of the utility, the following three wireless technologies are the primary ones used for this application (bridges between each of these wireless technologies and many different PLC technologies are available on the market).
   a. ZigBee
      ZigBee has been developed by the ZigBee Alliance and standardized by IEEE 802.15.4. It can operate in different frequencies (depending on the region) and between 20 and 250 kbit/s raw data rate. The ZigBee Alliance states that the transmission distance is between 10 and 1600 m. With the present application scenario, power consumption is critical, as batteries need to be used. Therefore, the transmission strength is reduced and, consequently, the lower end of the stated transmission distance range is more likely.

b. Wireless M-Bus

The Wireless M-Bus is specified in the European standard EN13757-4 and is mainly used for connecting meters in Europe, particularly Germany.

c. Z-Wave

Z-Wave has been specified by the Z-Wave Alliance and adopted as standard by the ITU (Recommendation ITU-T G.9959). It can operate in different frequencies depending on the region, and implementations provide different maximum data rates within the range of 9.6–100 kbit/s.

2.3.2 Convergent home networking

The previous Section 2.3.1 describes scenarios where a combination of wireless and PLC technologies is needed to transmit data from point A to point B. The supporting devices are basically bridges between technologies. This section describes the scenario where more than one physical path between point A and B is available. The paths are of different technologies, which form separate home networks. All available networks should converge for the optimization of the use of the existing infrastructure for multiple applications.

For example, some smart TV sets offer an Ethernet and a Wi-Fi 802.11 interface for Internet access. The user has the following options:

1. Install an Ethernet cable between the TV and the DSL router.
2. Use PLC between the DSL router and the TV's mains socket and then a short Ethernet cable between the PLC modem and TV.
3. Use Wi-Fi.

Option (1) requires the most effort and will therefore probably not be chosen. The user decides on option (2), because option (3), although available in the user's residence, does not provide good coverage at the location of the TV. At home, the user has three smart TVs. A system is needed that chooses automatically the best available communication path through the PLC and the Wi-Fi network by quality and amount of traffic on the link criteria. The decision might be revised if, for example, another TV is switched on.

IEEE introduced an "abstraction layer" to solve this technical issue and finished the standardization of this technology in 2013 with the standard 1905.1. Figure 2.3 shows the technologies that are supported by 1905.1.

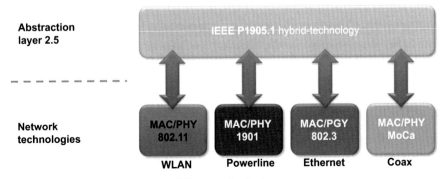

Figure 2.3 IEEE 1905.1 converges different technologies.

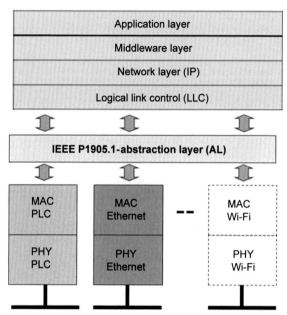

Figure 2.4 ISO open systems interconnection reference model.

The abstraction layer (also referred to as layer 2.5) is between the MAC layer of each supported technology and a common logical link control layer. It translates information between upper layer and different MAC/PHY technologies. Figure 2.4 shows the OSI layers.

Deciding on the best path(s) between devices supporting the abstraction layer as well as providing information about the topology and monitoring the path(s) is done by a managing software that is reachable either locally by the user PC or remotely by the Internet service provider's (ISP's) network management center. Although it is not clearly defined in the standard, this managing software is seen at present as a central intelligence in the resulting network – not as decentralized intelligence in every network node.

As the standardization has only been finished recently, products based on 1905.1 are currently under development.

2.4 Future trends

This section describes future trends in PLC and hybrid home networks in cases where first work has begun.

2.4.1 Improvements of the power line technologies

As described in the introduction, PLC has been continuously improved for more than 100 years since it has been invented. However, there is still room for further improvement.

The MIMO concept that has been described in Section 2.2.3 is seen as the key for higher coverage and higher data rates. The algorithms and the technical implementation in the chips are the subject of current research for improvement.

To make the best data performance out of a physical channel, OFDM improvements point to the direction of smaller bandwidth of a single subcarrier Δf (which results in a higher number of subcarriers) and higher modulation schemes.

For smart home application and building automation, where low data rates are acceptable, low prices for PLC nodes are essential. Therefore, low-cost implementations are the trend.

To avoid interference with radio services, the CENELEC EMC standard EN50561-1 for in-home PLC apparatus specifies dynamic frequency exclusion and automatic power management. Dynamic frequency exclusion is for the protection of broadcast radio services. If the field strength of the radio signal in a specific frequency and at the location of the radio broadcast receiver is above a defined minimum usable field strength level, the PLC apparatus excludes this frequency from PLC signal transmission. Automatic power management reduces the PLC power spectral density for transmission, depending on the channel quality, to a minimum. With this feature, the general EMC level is reduced on the one hand; on the other hand, the overall performance of a PLC home network with many PLC nodes is increased due to greater reusability of the PLC transmission spectrum by distant nodes – similar to the frequency reuse planning with mobile network base stations. Both features have just been recently developed or are currently under development by chip manufacturers, so that products offering them will access the market soon.

2.4.2 Improvements of hybrid networks

The convergence concept of home networks as specified by IEEE 1905.1 was described in Section 2.3.2. But IEEE 1905.1 only supports four technologies: Wi-Fi following IEEE 802.11, PLC compliant to IEEE 1901, Ethernet (IEEE 802.3), and Multimedia over Coax Alliance specification. Therefore, other technologies should be incorporated into the next specification. Potential candidates are ZigBee, Z-Wave, PLC (IEEE 1901.2), other PLC standards, and many more.

The present 1905.1 standard foresees a centralized point of intelligence (master) for the path decision, although it is not explicitly specified. For the next generation (e.g., version 2.0 or 1905.2, not yet named), efficient protocols and mechanisms for a decentralized intelligence approach are targeted. That means that each path can be negotiated between the two linked nodes.

With the increase in home network applications and the redundancy of home networks, more and more nodes are being installed in homes. This can potentially increase energy consumption. The obvious mitigation is to develop low-consuming devices; but in addition, low power consumption concepts on a network level are discussed. Protocols should be introduced to control power save modes and wake-up mechanisms of each node.

2.5 Sources of further information

In addition to the Webpages provided in the reference section, an excellent source of further information is the IEEE International Symposium on PLC and its Applications (ISPLC, http://www.ieee-isplc.org/). This symposium takes place annually and attracts PLC experts from all over the world.

- Loubery, C.R. (1899). *Einrichtung zur elektrischen Zeichengebung an die Theilnehmer eines Starkstromnetzes.* Berlin/Germany, Kaiserliches Patentamt, German Patent # 118717.
- CENELEC EN50065-1:2011, Signalling on low-voltage electrical installations in the frequency range 3–148.5 kHz – Part 1: General requirements, frequency bands and electromagnetic disturbances.
- CENELEC EN50561-1:2012, Power line communication apparatus used in low-voltage installations – Radio disturbance characteristics – Limits and methods of measurement – Part 1: Apparatus for in-home use.
- X10 Webpage. http://www.x10.com/productsupport/. Accessed 30.05.14.
- Evans, G. (2001). *CEBus Demystified: The ANSI/EIA 600 Users's guide.* McGraw-Hill.
- Echelon Webpage. http://www.echelon.com/technology/lonworks/. Accessed 30.05.14.
- PRIME Alliance Webpage. http://www.prime-alliance.org/. Accessed 20.08.13.
  - PRIME project: PRIME Technology White Paper. (July 21, 2008). Available at: http://www.prime-alliance.org/wp-content/uploads/2013/03/MAC_Spec_white_paper_1_0_080721.pdf. Accessed 20.08.13.
- Recommendation ITU-T G.9904. (2012). Narrowband orthogonal frequency division multiplexing power line communication transceivers for PRIME networks.
- G3 PLC Alliance Webpage. http://www.g3-plc.com/. Accessed 20.08.13.
  - ERDF: PLC G3 Physical Layer Specification. (October 08, 2009).
  - Maxim: Supplement to PLC G3 physical layer specifications for operation in the FCC frequency band. (November 12, 2010)
  - Jean Vigneron. General Secretary of G3-PLC Alliance, and Kaveh Razazian, Senior Scientist – Maxim Integrated, presentation "G3-PLC Power line Communication Standard for Today's Smart Grid." (October 2012).
- Recommendation ITU-T G.9903. (2012). Narrowband orthogonal frequency division multiplexing power line communication transceivers for G3-PLC networks.
- Michael Koch. High speed narrowband PLC field trials. (June 20, 2012). Available at: http://www.dlansolutions.de/wp-content/uploads/2012/10/2012-07-19_SG-Paris_devolo_Dr-Michael-Koch.pdf. Accessed 20.08.13.
- IEEE 1901.2. 2013. Standard for Low Frequency (less than 500 kHz) Narrow Band Power Line Communications for Smart Grid Applications. ISBN: 978-0-7381-8794-5. 2013.
- HomePlug Powerline Alliance Webpage. https://www.homeplug.org/home/. Accessed 20.08.13.
- IEEE 1901. "Standard for Broadband over Power Line Networks: Medium Access Control and Physical Layer Specifications," project Webpage at http://grouper.ieee.org/groups/1901/. Accessed 21.08.13.
- ISO/IEC Joint Working Group on ISO project-Nr. 15118-3: Road vehicles – Vehicle to grid Communication Interface – Part 3: Physical and data link layer requirements. Draft is in the standard approval process.
- ZigBee Alliance Webpage. http://www.zigbee.org/Home.aspx. Accessed 22.08.13.
- Z-Wave Alliance Webpage. http://www.z-wavealliance.org/. Accessed 22.08.13.
- IEEE 1905.1. "Standard for a Convergent Digital Home Network for Heterogeneous Technologies," project Webpage at http://grouper.ieee.org/groups/1905/1/. Accessed 22.08.13.

Middleware platform for smart home networks

N. Ishikawa
Komazawa University, Setagaya, Tokyo, Japan

3.1 Standardization of home networks and devices

Much standardization has been done for controlling and managing home networks and devices during the past 10 years. Such activities include Universal Plug and Play (UPnP)/Digital Living Network Alliance (DLNA), Energy Conservation and Homecare Network (ECHONET), Continua Health Alliance, and so on. We briefly describe such standardization activities on home network devices, and the current status on development and deployment of products based on such standards.

3.1.1 Universal Plug and Play/Digital Living Network Alliance

UPnP (UPnP Forum) is a technology that allows the use of standard Internet technology for connections between networked devices. The DLNA has developed design guidelines for the sharing of video and other digital content by home appliances, personal computers, and mobile devices through the use of UPnP and other standard Internet technology. Digital media players (DMP), which present content for user enjoyment, use protocols such as Simple Service Discovery Protocol (SSDP) to discover digital media servers (DMS) that have contents, and to retrieve a list of contents from a content directory service. Then a request for the desired content is sent and the content is streamed from the DMS via the local area network to the DMP for viewing.

The DLNA Guidelines 1.0 and 1.5 were developed in October 2004 and March 2006, respectively, and DLNA-compliant devices have been developed, focusing on PCs, TVs, hard disk drives (HDD) and Blu-ray disk players, digital music players, and so on. DLNA is also conducting a logo certification program to ensure interconnectivity between products developed by various manufacturers. To protect copyright in the transfer of content, the Link Protection Guidelines for Digital Transmission Content Protection over IP (DTCP-IP) were developed in October 2006. UPnP/DLNA-compliant audiovisual (AV) devices have been developed and deployed to allow sharing of downloaded digital content (e.g., video, audio) among AV devices, listening to digital music from HDD via a digital music player, and so on.

3.1.2 Energy conservation and homecare network

ECHONET (ECHONET CONSORTIUM) is a standard for control of air conditioners, lighting equipment, and other home appliances and facility devices such as power consumption monitors and other sensors.

Ecological Design of Smart Home Networks. http://dx.doi.org/10.1016/B978-1-78242-119-1.00003-5

While version 1.0 of the ECHONET specifications was developed in March 2000, some manufacturers have developed ECHONET-compliant products, and such products are available in the market, but are generally not widely available. Because the recent trend is to communicate with AV devices, version 3.60 of the ECHONET specifications was published in December 2007. It includes gateway specifications that allow conversion between ECHONET and UPnP.

ECHONET Lite was published in December 2011. ECHONET Lite simplifies ECHONET specifications by removing their specifications in the physical layer. The main purpose of ECHONET Lite is to apply ECHONET specifications to home energy management systems (HEMS). For this purpose, ECHONET Lite defines an ECHONET object for a smart meter, and ECHONET properties for energy management on home appliances such as air conditioners.

3.1.3   Open Services Gateway Initiative

Open Services Gateway Initiative (OSGi) (OSGi Alliance) was founded in March 1999 and has since been working on standards for mechanisms for the remote and dynamic addition and changing of home gateway (HGW) functions. OSGi middleware is installed on HGWs that execute Java Virtual Machine (VM). OSGi middleware enables an HGW to add various functions to be executed on it, by downloading specific Java components called bundles from cloud servers. OSGi release 4, published in 2005, extended the OSGi specifications for mobile devices. OSGi middleware has been implemented and used on various devices such as mobile phones and in-vehicle devices, in addition to HGWs.

3.1.4   Continua Health Alliance

Continua Health Alliance was founded in June 2006. Continua Health care design guidelines version 1 was developed in June 2009. The guidelines define specifications that are based on ISO/IEC 11073 Personal Health Data (PHD) standards (ISO TC 215 Health informatics), between health care devices and gateways (e.g., PC) via USB and Bluetooth. Such health care devices include blood pressure monitors, thermometers, weighing scales, and so on. Many health care devices that are available in the market are compliant with Continua Health Alliance design guidelines. PC software that implements protocol stacks and device drivers, which are compliant with the above design guidelines, are available from some software vendors.

3.2   Problems with standardization regarding home networks and devices

Various standards have been developed for different classes of home devices as described in Section 3.1. There is no compatibility among those standards. In addition, there are no standards for security devices such as security cameras and electronic

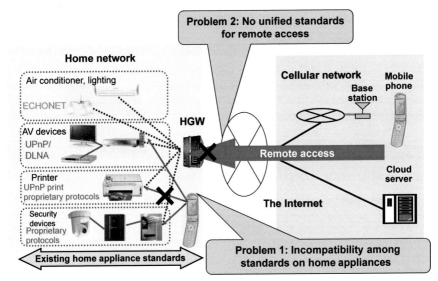

Figure 3.1 Problems regarding remote access to home networks.

keys. Such devices support proprietary specification developed by vendors. Two main problems posed by this situation are described below (Figure 3.1).

Problem 1: Different classes of home devices support different standards.

The current situation is that there is no compatibility among the various standards. A mobile phone to control home devices compliant with different standards must support multiple standards. However, current mobile phones cannot support many standards for home devices, due to limited processing capability and memory capacity. In addition, HGWs must also support many standards to control and manage home devices that comply with different standards.

Problem 2: No standards for controlling and managing home devices from outside the home.

From the mobile phone point of view, it is very important for a mobile phone to be able to remotely control home devices via a cellular network. From the cloud server point of view, it is very important for a cloud server to be able to remotely manage home devices via the Internet. However, many standards on home devices only define protocols for communications within a home network. In current standards on home devices, little attention has been paid to communications for remote control and management of home devices from outside the home.

To solve those problems, we have conducted joint research with Ericsson, Kyoto University, and Keio University to develop peer-to-peer/overlay networking and metadata technologies as a solution for unified control and management of various home devices connected to home networks, from mobile phones and cloud servers. Based on the results of this joint research, we established the Peer-to-peer Universal Computing Consortium (PUCC) in 2005, to deploy our overlay networking and metadata technologies toward de facto standardization. Such activities include the development of PUCC specifications, development of prototype systems using PUCC technologies, demonstrations on PUCC technologies at world-famous exhibitions (e.g., Consumer Electronics Show (CES) and Mobile World Congress (MWC)) and the efforts toward International Standardization (Ishikawa, 2013).

In this chapter, we describe our peer-to-peer/overlay networking and metadata technologies, and the general middleware platform, which have implemented such technologies in more detail. Our middleware platform is now running over PCs, HGWs, mobile phones including feature phones and smart phones, tablet devices, Web servers, and so on. Some applications that we have developed over our middleware platform are also described.

3.3 PUCC peer-to-peer/overlay networking protocols and device metadata technologies

Compared with traditional home networking technologies described in Section 3.2, peer-to-peer/overlay networking technologies have functions to realize resource discovery, multihop networking in a highly distributed manner. Peer-to-peer/overlay networking is one of the most important and suitable technologies for communications among home devices over heterogeneous home networks. In addition, metadata is also an important technology for describing various home devices in a consistent manner.

To this purpose, we have developed peer-to-peer/overlay networking protocols and metadata technologies, initially developed by joint research among NTT DoCoMo, Ericsson, and Kyoto University. We also established PUCC to deploy our overlay networking protocols and device metadata technologies toward de facto standardization. The latest PUCC specifications Release 3, from March 2012, defines 11 specifications as shown in Figure 3.2. PUCC activities are described in detail in Ishikawa (2013). PUCC peer-to-peer/overlay networking protocols and device metadata technologies are described below (Ishikawa, Kato, Sumino, Murakami, & Hjelm, 2007).

3.3.1 PUCC architecture

In the PUCC architecture, bidirectional communication entities, called PUCC nodes, construct a PUCC network by establishing a PUCC session between them. The PUCC nodes communicate with each other using the PUCC sessions. Each PUCC node has a unique ID. The key elements of the PUCC architecture are defined as follows.

PUCC node: PUCC node is an independent, bidirectional communication entity. In the PUCC architecture, it can be PCs, HGWs, home appliances, mobile phones, tablet devices, Web servers, or any of a variety of devices. Each node has a unique ID and communicates using the ID independent from underlying networks (e.g., the Internet).

PUCC network: The term *PUCC network* means a logical collection of PUCC nodes that have a common interest and obey a common set of policies. A PUCC session between PUCC nodes is established on mutual trust. Each PUCC node can enter or depart the PUCC network at its convenience. PUCC messages are sent from one PUCC node to another directly or via some intermediary PUCC nodes. Routing information is discovered by broadcasting an inquiry message to the PUCC network.

(1) **PUCC Architecture Ver. 2**
(2) **PUCC Basic Protocol Ver. 3**
(3) **PUCC Basic Protocol - Light Profile Ver. 2**
(4) **PUCC Device Discovery Service Invocation Protocol Ver. 3**
(5) **PUCC Printing Protocol Ver. 3**
(6) **PUCC Imaging Protocol Ver. 2**
(7) **PUCC Streaming Protocol Ver. 3**
(8) **PUCC Device and Service Metadata Template Ver. 3**
(9) **PUCC Metadata Specification –**
   - **Part1: IEEE 1394 Devices Ver. 1**
   - **Part2: UPnP Devices Ver. 2**
   - **Part3: ECHONET Devices Ver. 1**
   - **Part4: Other Devices Ver. 1**
   - **Part5: Camera Device Ver. 1**
   - **Part6: RFID Reader Device Ver. 1**
   - **Part7: Felica Reader Device Ver. 1**
   - **Part8: Barcode Reader Device Ver. 1**
   - **Part9: Location Information Ver. 1**
   - **Part10: IEEE 11073 Device Ver. 1**
   - **Part11: OSGibased Device Ver. 1**

Figure 3.2 PUCC specifications Release 3.

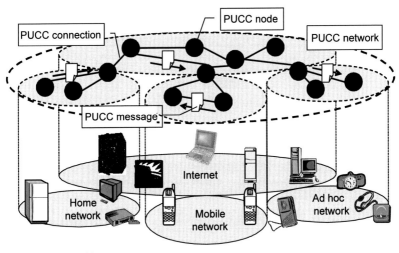

Figure 3.3 PUCC architecture.

PUCC message: This is data that is sent and received between PUCC nodes. A PUCC message is a basic unit of exchanging data and has a unique ID.

PUCC session: This is a communication channel established between PUCC nodes. PUCC messages are transmitted along the PUCC sessions.

The PUCC architecture is shown in Figure 3.3.

In the PUCC communication model, the role of each communicating entity is not always clearly distinguishable. In order to design effective PUCC communication

Figure 3.4 PUCC communication model.

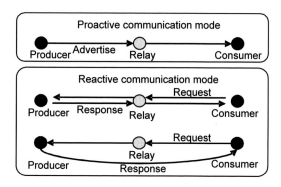

protocols, the "role" concept is introduced. We have established the PUCC communication model by looking into existing peer-to-peer applications. For example, we found three kinds of nodes in peer-to-peer distributed search applications (Clarke, Sandberg, Wiley, & Hong, 2000; Liang, Kumar, & Ross; The Gnutella protocol specification v0.4). The first provides content information (e.g., its location) in response to search requests. The second requests the content search. The third relays the search request and their replies. Those applications suggest that PUCC nodes can have three roles:

Producer role: A node acts as a producer when it provides application data or services.

Consumer role: A node acts as a consumer when it asks for a service or consumes application data without any request or in response to its request.

Relay role: A node acts as a relay when its current communication task is to forward application data, service requests, and their replies.

In peer-to-peer applications, a node may play any of these three defined roles, and the peer-to-peer application dynamically determines which one is to be used. Based on the roles described above, we define two PUCC communication modes (Figure 3.4):

Proactive communication mode: This mode represents the unsolicited transmission of information that does not require any specific response. It is generally used by a PUCC node to notify other PUCC nodes of its own existence or resources it holds.

Reactive communication mode: This mode represents the transmission of information that requires a response. It is generally used by a PUCC node that requests certain services or resources provided by other PUCC nodes.

As shown in Figure 3.5, the PUCC architecture also supports basic communication types such as unicast (includes multihop unicast), broadcast, and multicast.

In PUCC networks, each PUCC node should have a unique name. This name will be extensively used for various purposes such as node search, routing information to a node, and cache table management. PUCC node names should satisfy the following requirements.

Uniqueness: PUCC node names should be unique within the scope of a community.

Manageability, simplicity: PUCC node names should be able to be autonomously generated in particular for pure ad hoc PUCC networks. Its naming structure should be simple for manageability and generality.

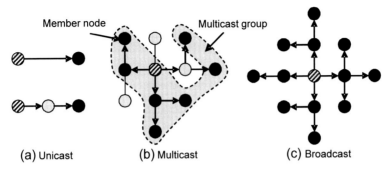

Figure 3.5 PUCC communication types.

Scalability: The PUCC naming system should support an extremely large number of nodes.

Anonymity, privacy, and security: Nobody should be able to infer any type of private information from PUCC node names. For security, PUCC node names should not be arrogated.

Independence: PUCC node names should be independent of location, user, transport protocol, application, and so on.

Considering the above requirements, we have defined PUCC node names that are based on UUID (Leach, Mealling, & Salz, 2005) where a node name is assigned by the node itself. UUID is, however, not human readable, and an alias name system for human readability may be required.

Message routing is one of the key mechanisms for realizing efficient and reliable peer-to-peer/overlay communication. With the traditional Internet, routing is performed by a router according to a routing table it holds. However, since a PUCC node freely enters and leaves a PUCC network, the topology of a PUCC network changes very frequently. Routing based on a stable routing table is hence inadequate and inefficient for a PUCC network. Since PUCC nodes communicate with each other across heterogeneous network environments, a transport layer independent routing mechanism is required. In the PUCC architecture, a name-based routing mechanism is adopted. This is a mechanism that finds a source route toward a destination PUCC node in a heuristic manner. Good examples of this type of routing mechanism are ad hoc network routing mechanisms such as dynamic source routing (DSR; Johnson, Hu, & Malz, 2007). One advantage of this method is its simplicity because it does not need complicated routing protocols and a particular server. Obviously, it consumes many network resources and incurs long delay. Hence, performance and efficiency could not be sufficiently ensured in large PUCC network environments.

3.3.2 PUCC protocols

The PUCC protocols are designed to realize the PUCC architecture. The PUCC protocols described herein are designed according to the following requirements:

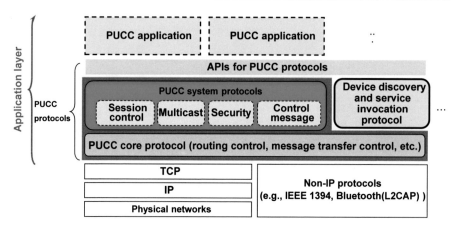

L2CAP: Logical link control and adaptation protocol

Figure 3.6 Architecture for PUCC protocols.

Extensibility: PUCC protocols should be layered, generic, and have extensibility so that they can support various peer-to-peer applications.

Utilization of existing technology: PUCC protocols should leverage existing technologies such as XML, and support existing network infrastructures such as the Internet to simplify implementation and deployment.

Independence of transport protocols: PUCC protocols should be independent of transport protocols for realizing peer-to-peer applications over heterogeneous network environments (e.g., the Internet, home networks, and ad hoc networks).

As shown in Figure 3.6, the PUCC protocols are currently defined over TCP/IP and non-IP protocols such as Bluetooth (Bluetooth), IEEE 1394 (IEEE), and OBEX/IrDA (Infrared Data Association). PUCC protocols are defined as overlay networking protocols in the application layer. PUCC protocols hence allow communication among home devices, HGWs, mobile phones, and cloud servers across heterogeneous networks.

PUCC protocols are application-independent and general-purpose protocols for various PUCC applications, which include home devices controlled from a mobile phone, multicast video streaming, sensor network applications, and so on. The application programming interfaces (APIs) for PUCC protocols are defined for developing various PUCC applications. The PUCC protocols stack comprises the two layers of the PUCC core protocol and the other PUCC protocols over the PUCC core protocol as shown in Figure 3.6. The PUCC protocols are briefly described below.

1) PUCC core protocol

This protocol processes PUCC messages according to the PUCC communication model. Three message types are defined in Section 3.3.1. Request and response messages are defined for the reactive communication mode, while an advertise message is defined for the proactive communication mode. A message is sent to the destination node either directly or using multihop unicast. A node sends a broadcast message to all adjacent nodes. The message routing mechanisms of the PUCC core protocol are independent of underlying transport protocols. Other PUCC protocols are defined over this protocol.

2) PUCC system protocols

PUCC system protocols define common protocols for various PUCC applications, which include the PUCC basic communication protocol, the PUCC multicast communication protocol, and the PUCC control message protocol.

The PUCC basic communication protocol establishes and releases PUCC sessions. In the PUCC architecture, all communications are based on a PUCC session between pair adjacent PUCC nodes. This protocol also has the function of exchanging node resource information such as names of its adjacent PUCC nodes, and security functions such as encryption and authentication.

The PUCC multicast communication protocol constructs a multicast distribution tree for multicast member nodes and forwards multicast messages along it. A node finds a member node of a multicast group and sends a Join message to it, and a multicast routing table is generated at each node along the path toward it at the same time. When a node wants to send a multicast message, it sends the message toward the adjacent member nodes based on the multicast routing table. The multicast messages are forwarded along the multicast distribution tree using the bidirectional shared tree mechanism. When the node leaves the multicast group, it sends a Leave message to the adjacent member nodes. This protocol is mainly used by multicast streaming applications.

The PUCC control message protocol provides ancillary functions such as notification of message forwarding error, keep-alive for PUCC sessions, and first PUCC node discovery in PUCC network environments. For example, an ErrorReport message is used to notify the source node of the forwarding error of a message. A Diagnose message is used to measure round trip time between PUCC nodes. A Lookfor message is used to find the first peer node to which a node should connect in PUCC network environments.

3) PUCC application protocols

In addition to PUCC system protocols, PUCC application protocols can be defined over the PUCC core protocol for various PUCC applications. For smart home applications, the PUCC device discovery and service invocation protocol is the most important PUCC application protocol. This protocol is described below in more detail.

To control and manage home devices that are connected to a home network, it is first necessary to discover each home device, its capabilities, and the services offered by it. The overview of this protocol is shown in Figure 3.7. We define a PUCC general-purpose device metadata template. The device metadata for each device is defined using this device metadata template. Such devices include home appliances, health care devices, surveillance cameras, sensor devices, and so on.

This protocol uses PUCC device metadata defined for various devices to discover and control devices and the services they offer, as shown in Figure 3.7(a). Each device on a home network has metadata that describes its capabilities and the services it offers. A user searches for home devices by broadcasting a PUCC search request message

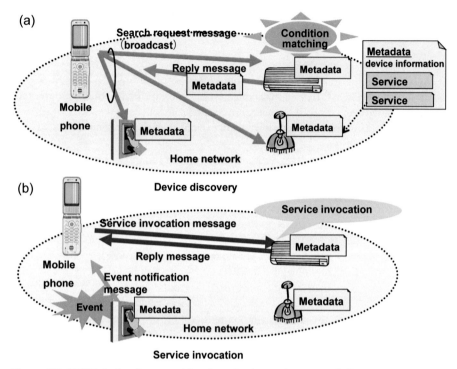

Figure 3.7 PUCC device discovery (a) and service invocation protocol (b).

over the PUCC overlay network, which contains device functions, device types, or key words as search conditions. Each device that receives the PUCC search request message compares its own device metadata with the search conditions and returns a reply message that contains the device metadata if the description included in the device metadata matches the search conditions, as shown in Figure 3.7(b). The user selects the service to be invoked from the services described in the received device metadata and then sends the PUCC service invocation message to control the device. Since a mobile phone automatically creates a graphical user interface (GUI) for controlling home devices, based on the description included in the device metadata, a user easily controls a home device by using the GUI generated on the display of the mobile phone.

In addition, the PUCC device discovery and service invocation protocol defines the subscription and notification functions. For example, a cloud server subscribes to a home device such as a health care device, using the PUCC subscription message. If an event (e.g., measuring health care data from a health care device or sensor data from a sensor device) occurs, the device notifies the cloud server of the event, using the PUCC notification message. Event conditions (e.g., notification of an event once every hour) can be also described in the metadata of a device. The PUCC subscription and notification functions are applied to communications between home devices connected to a home network and a cloud server, via an HGW. Those functions may be applied to various PUCC applications, including health care applications and HEMS.

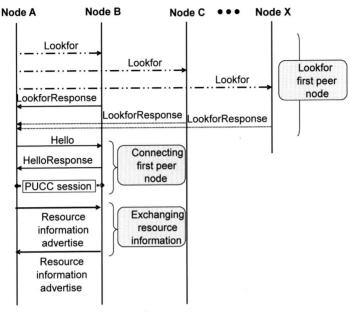

Figure 3.8 Example of a PUCC protocol sequence.

Figure 3.8 shows a basic sequence when a PUCC node participates in a PUCC network and exchanges resource information with a first peer node in the PUCC network. At first, Node A sends a Lookfor message using a network-specific broadcast or multicast mechanisms (e.g., IP multicast) and receives the corresponding Lookfor-Response messages from certain nodes. Then, Node A sends a Hello message to one of the discovered PUCC nodes (in this example, Node B) to participate in a PUCC network. When a HelloResponse message from Node B is received, a PUCC session is established between Node A and Node B. Resource information is exchanged using resource information advertisement messages in the next step.

3.3.3 PUCC device metadata

The structure of the PUCC device metadata is shown in Figure 3.9. The PUCC device metadata describes information on the device including device name, type, device attributes, and the services the device offers, using XML format.

Static information on a device is described in the "specification" field, which includes device name, device manufacturer, model number of the device, serial number of the device, and universal product code. The various functions provided by the device are described as a list of services in the "Service List" field, which include function name, input parameters, and output parameters. When defining a service, it is possible to specify the service as a series of functions. For example, a "view DVD" service could be specified as the following series of functions. Each service has a unique uniform resource identifier (URI), and the URI is used as an identifier for service discovery and service invocation.

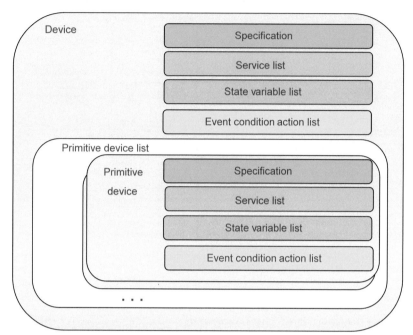

Figure 3.9 Structure of PUCC device metadata.

1. Turn on a TV
2. Set input of the TV to "video"
3. Turn on a DVD player.

The "State Variable List" field is used to define a list of state variables. Each state variable holds a value that a device measures. For example, a state variable named "temperature" is used as a state variable in a thermometer device metadata, to hold the value of temperature that the thermometer measures.

The "Event Condition Action List" field can be used to define a list of event–condition–action (ECA) rules among devices. An ECA rule has three parts: an event, a condition, and an action. The semantics of an ECA rule are: when the event is detected, the condition is evaluated, and if the condition is satisfied, the action will be executed. For example, using this field, the following ECA rules can be defined:

- When a thermometer shows over 28 °C, turn on an air conditioner.
- When a thermometer shows below 24 °C, turn off an air conditioner.

The "Primitive Device List" field is used for defining individual components in a multifunction device. For example, a Blu-ray recorder device may be defined as a multifunction device. In this case, metadata for a Blu-ray player, a Blu-ray recorder, and an HDD recorder are defined in each "Primitive Device" field included in the "Primitive Device List" field.

The PUCC device metadata is compatible with UPnP metadata to ensure interoperability with UPnP-compliant devices, and has been designed as general-purpose

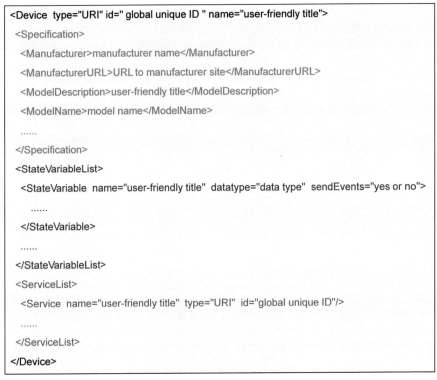

```
<Device type="URI" id=" global unique ID " name="user-friendly title">

  <Specification>

  <Manufacturer>manufacturer name</Manufacturer>

  <ManufacturerURL>URL to manufacturer site</ManufacturerURL>

  <ModelDescription>user-friendly title</ModelDescription>

  <ModelName>model name</ModelName>

  ......

  </Specification>

  <StateVariableList>

  <StateVariable name="user-friendly title" datatype="data type" sendEvents="yes or no">

      ......

  </StateVariable>

  ......

  </StateVariableList>

  <ServiceList>

  <Service name="user-friendly title" type="URI" id="global unique ID"/>

  ......

  </ServiceList>

</Device>
```

Figure 3.10 Framework of PUCC device metadata description.

device metadata by enabling it to be applied to metadata description for other devices including sensor devices and health care devices, and adding functions for describing ECA rules.

To interwork with home devices that are compliant with existing standards on home devices (e.g., UPnP/DLNA and ECHONET), PUCC device metadata are defined to be compatible with such standards. When controlling such devices, an HGW converts the PUCC device discovery and service invocation protocol into existing standard protocols for controlling home devices.

Figure 3.10 shows a framework for PUCC device metadata description using XML.

3.4   PUCC middleware platform

We have developed the PUCC middleware platform based on the PUCC specifications described. The requirements are as follows:

1. The PUCC middleware platform should support a wide range of OS environments.
2. The PUCC middleware platform should run over a wide range of devices.
3. The PUCC middleware platform should provide general-purpose APIs for software developers who will implement various PUCC applications.

Considering the above requirements, we have developed the PUCC middleware platform over Java 2 SE and Java 2 ME environments. As shown in Figure 3.11, the main components of the PUCC middleware platform are the PUCC protocols module, the PUCC node manager, and the transport adapter. The PUCC protocols module implements PUCC protocols including the PUCC core protocol, PUCC systems protocols, and PUCC application protocols. The role of the PUCC node manager is the management of PUCC nodes (e.g., invocations and termination of PUCC nodes) on a device. The role of the PUCC transport adapter is to map PUCC protocols onto various underplaying transport protocols (e.g., TCP/IP, Bluetooth, IEEE 1394, and OBEX/IrDA).

Since the PUCC middleware platform runs over Java 2 SE environments, it works on various operating systems including Microsoft Windows, Unix/Linux, and Apple Mac OS. Since the PUCC middleware platform runs over Java 2 ME environments, it works on various small and embedded devices including mobile phones. For example, we have implemented the PUCC middleware platform as a Java application, called i-Appli, on NTT DoCoMo's feature phones (Sumino, Ishikawa, & Kato, 2006). i-Appli can be developed over Java 2 ME CLDC environments, which are supported by NTT DoCoMo's feature phones.

We have also converted the PUCC middleware platform into Java format called "OSGi bundle" specified in OSGi Alliance. The PUCC middleware platform then runs over OSGi-compliant HGWs. We have converted the PUCC middleware platform into Android software, which runs over Dalvik VM, provided by Google. The PUCC middleware platform then runs over Android devices including smartphones and tablets.

The PUCC middleware now runs over a wide range of devices including PCs, HGWs, mobile phones, tablet devices, and cloud servers.

We have also developed Java APIs for PUCC application developers. Java APIs roughly consist of PUCC protocols APIs and PUCC management APIs. PUCC protocols APIs provide PUCC application developers with functions for using PUCC protocols including the PUCC core protocol, PUCC system protocols, and PUCC application protocols (e.g., PUCC device discovery and service invocation protocol). PUCC application developers also design their own PUCC application protocols, using APIs for the PUCC core protocol. PUCC management APIs provide PUCC application developers with management functions such as invocation and termination of PUCC nodes on a device, and error handling.

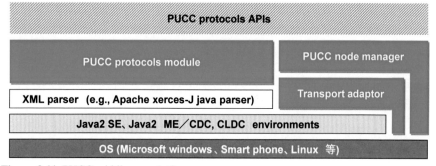

Figure 3.11 PUCC middleware platform.

3.5 Smart home applications using the PUCC middleware platform

To verify the feasibility and usefulness of PUCC technologies, PUCC has developed prototype systems implementing PUCC middleware platforms for mobile phones, PCs, tablet devices, HGWs, cloud servers, and so on. PUCC has also developed PUCC applications such as multimedia content search from mobile phones, home appliances control from mobile phones, health care applications, AV devices (e.g., HD/DVD recorders) control from mobile phones, and home security applications. PUCC conducted several demonstrations on PUCC technologies at world-famous exhibitions such as CES 2008, CES 2009, MWC 2009, and CeBIT 2009 (Ishikawa, 2013). Among them, we described several smart home applications using PUCC middleware platforms.

3.5.1 Security camera control from mobile phones

Home security application is one of the most important applications for smart homes. Such applications include security camera control and front door control (i.e., checking whether a front door is locked) from mobile phones outside the home. Among them, we describe the security camera control system from mobile phones (Ishikawa, Kato, & Osano, 2011). The configuration of the security camera system using mobile phones is shown in Figure 3.12. This system consists of high-definition security cameras, a camera-control gateway (GW), and mobile phones. The system transfers high-definition H.264 video from the surveillance cameras to the GW. The GW converts from H.264 video to motion JPEG video, and transfers it to mobile phones. The GW converts H.264 video to motion JPEG in real time. The PUCC protocols are used to control the security cameras (i.e., pan, tilt, and zoom). To control devices, the PUCC protocols use device metadata written

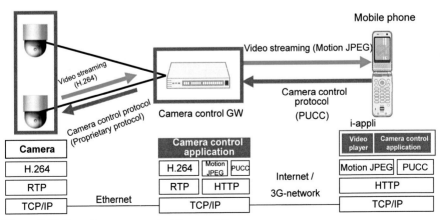

Figure 3.12 Security camera control from mobile phones.

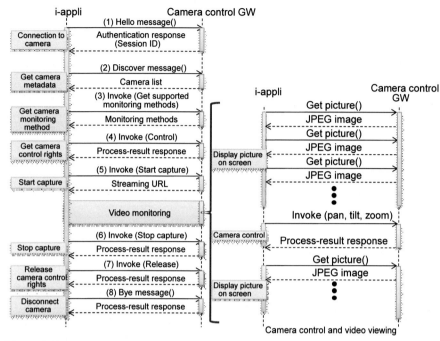

Figure 3.13 PUCC protocol sequence for security camera control.

in eXtensible Markup Language (XML) describing device name, type, attributes, and provided services. PUCC metadata for camera devices are defined in PUCC specifications Release 3 (Figure 3.2). Using PUCC protocols in this way enables functional differences in each camera (e.g., zoom control can be performed but pan/tilt control cannot) to be described by device metadata, which means, in turn, that cameras with a variety of functions can be manipulated by the same PUCC application on a mobile phone.

The PUCC control sequence is shown in Figure 3.13 and summarized below.

1. Perform connection processing and authentication from the i-appli on the mobile phone side to the GW and get SessionID (Hello message).
2. Get metadata for camera devices targeted for connection (Discover message). This action enables a list of cameras to be obtained.
3. Get information on the video delivery capabilities supported by each camera (send a Get-SupportedMonitoringMethods command by the Invoke message).
4. Get camera usage right for camera device selected by the user (send a Control command by the Invoke message).
5. Request video capture to begin (send a StartCapture command by the Invoke message). The GW returns a URL as a response to the Invoke message and the i-appli requests the GW to get Motion JPEG with respect to that URL. This action starts the GW transferring the camera video and displays it on the mobile phone. The camera can be controlled at this time by camera-control requests (i.e., by sending Pan, Tilt, and Zoom commands using the Invoke message).

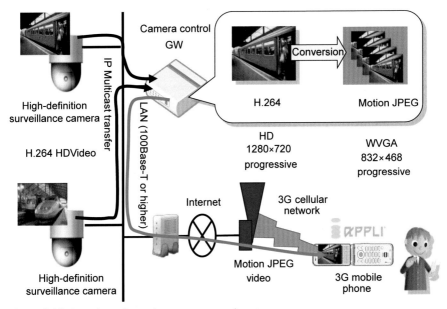

Figure 3.14 Overview of security camera control system.

6. Request video-capture stop-processing to terminate viewing of camera video (send a StopCapture command by the Invoke message).
7. Release camera usage rights (send a Release command by the Invoke message).
8. Perform disconnect processing from the i-appli on the mobile phone side to the GW (Bye message).

An outline of this system is shown in Figure 3.14. The cameras used here are surveillance cameras capable of streaming H.264 high-definition video with an output picture size of 1280×720 pix. The application on the mobile phone side was implemented as an i-appli. The GW incorporates an application for converting 1280×720 pix H.264 high-definition video to WVGA-equivalent 832×468 pix motion JPEG video in real time. Although this system uses high-end surveillance cameras as research, the same system configuration can be applied to the home security camera control system using cheaper security cameras.

3.5.2 Home appliances control from smart devices

We have also developed the home appliances (e.g., air conditioner) control system from smart devices (e.g., smart phone and tablet device).

Much attention has been paid to ECHONET Lite specifications as standards for smart home applications in Japan. The main purpose of ECHONET Lite is to apply ECHONET specifications to HEMS. Some ECHONET Lite-compliant commercial products are now available in Japan. In this system, we use ECHONET Lite-compliant air conditioners and lamps. OSGi-compliant commercial HGWs are also available. Using those devices, we have developed ECHONET

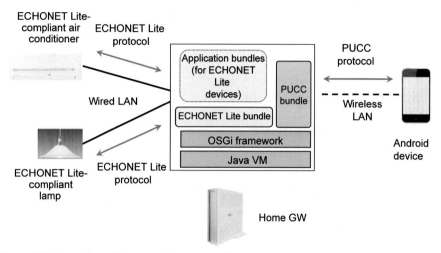

Figure 3.15 Overview of home appliances control system.

Lite-compliant home appliances from smart devices. Since we have already implemented the PUCC middleware platform on Android devices, we use Android smart phones and Android tablets in this system.

An overview of this system is shown in Figure 3.15. This system consists of smart devices, HGWs, and ECHONET Lite-compliant air conditioners and lamps. This system works as follows:

1. OSGi bundles including the PUCC bundle can be downloaded from a management center, if necessary.
2. A smart device (i.e., Android smart phone or Android tablet) looks for PUCC devices using a Lookfor message, and finds an HGW. The PUCC session is then established between the smart device and the HGW, using Hello messages. The smart device sends a Discover request message to the HGW, to search ECHONET Lite-compliant devices.
3. When the HGW has received the Discover request message, the application bundle for ECHONET Lite devices converts it to the corresponding ECHONET Lite protocol using the ECHONET Lite bundle, to search ECHONET Lite-compliant devices. When ECHONET Lite-compliant devices are discovered (in this case, air conditioners and lamps), the HGW sends a Discover reply message to the smart device, which contains ECHONET Lite device metadata for air conditioners and lamps.
4. When the smart device receives the Discover reply message, it generates the GUI for controlling ECHONET Lite-compliant devices, using ECHONET Lite device metadata contained in it. When a user wants to control ECHONET Lite-compliant devices, the user can control them using the GUI on the smart device. For example, when the user wants to turn on the air conditioner, the smart device sends an Invoke request message, which contains the service request (i.e., turn on the air conditioner) to the HGW. When the HGW receives it, the application bundle for ECHONET Lite devices converts it to the corresponding ECHONET Lite protocol using the ECHONET Lite bundle to turn on the air conditioner. As a result, the air conditioner is turned on.

3.6 Conclusions

We first described the current trend in standardization on home networks and devices, and then issues with such standardization activities from the mobile phone and cloud server point of view. To resolve such issues, we have developed PUCC overlay networking protocols and device metadata technologies as a solution for unified control and management of various home devices connected to home networks, from a mobile phone and a cloud server.

We have developed the PUCC middleware platform, based on the PUCC specifications. Since the PUCC middleware platform runs over Java 2 SE environments, it works on various operating systems including Microsoft Windows, Unix/Linux, and Apple Mac OS. The PUCC middleware now runs over a wide range of devices including PCs, HGWs, mobile phones, tablet devices, and cloud servers. To verify the feasibility and usefulness of PUCC technologies, PUCC developed several PUCC applications, including smart home applications.

Future work will include further development of PUCC applications such as HEMS, applications for smart grid, health care applications, distributed autonomous control of home devices and M2M applications toward realizing smart home and smart city, integration of PUCC technologies with social networking services (e.g., Facebook), and further efforts toward international standardization.

Acknowledgements

This work was supported by NTT DOCOMO, Ericsson, Kyoto University, Keio University, and PUCC.

This work was partially supported by the Special Research Program of Komazawa University, 2013.

References

Bluetooth, http://www.bluetooth.com/Pages/Bluetooth-Home.aspx.

Clarke, I., Sandberg, O., Wiley, B., & Hong, T. W. (2000). Freenet: a distributed anonymous information storage and retrieval system. In *Proceedings of the ICSI workshop on design issues in anonymity and unobservability*. California: Berkeley.

Continua Health Alliance, http://www.continuaalliance.org/index.html.

DLNA, http://www.dlna.org/.

ECHONET CONSORTIUM, http://www.echonet.gr.jp/english/index.htm.

IEEE 1394, http://standards.ieee.org/develop/wg/1394_WG.html.

Infrared Data Association, http://www.irda.org/.

Ishikawa, N. (2013). PUCC activities on overlay networking protocols and metadata for controlling and managing home networks and appliances. *Proceedings of the IEEE, 101*(11), 2355–2366.

Ishikawa, N., Kato, T., & Osano, T. (2011). High-definition surveillance camera control system from mobile phones. In *8th IEEE consumer communications and networking conference (CCNC 2011)*.

Ishikawa, N., Kato, T., Sumino, H., Murakami, S., & Hjelm, J. (2007). PUCC architecture, protocols and applications. In *4th IEEE consumer communications and networking conference (CCNC 2007)*.

ISO TC 215 Health informatics, http://www.iso.org/iso/home/standards_development/list_of_iso_technical_committees/iso_technical_committee.htm?commid=54960.

Johnson, D., Hu, Y., & Malz, D. (2007). *The dynamic source routing protocol (DSR) for mobile ad hoc networks for IPv4* RFC 4728.

Leach, P., Mealling, M., & Salz, R. (2005). *A Universally Unique IDentifier (UUID) URN Namespace* RFC 4122 .

Liang, J., Kumar, R., & Ross, K. W. The KaZaA Overlay: A Measurement Study. http://www.di.unipi.it/~ricci/TutorialKazaa.pdf.

OSGi Alliance, http://www.osgi.org/Main/HomePage.

Sumino, H., Ishikawa, N., & Kato, T. (2006). Design and implementation of P2P protocol for mobile phones. In *3rd IEEE workshop on mobile peer-to-peer computing (MP2P06), IEEE PerCom*.

The Gnutella protocol specification v0.4, http://web.stanford.edu/class/cs244b/gnutella_protocol_0.4.pdf.

UPnP Forum, http://www.upnp.org/.

Part Two

Applications

The ecology of home sensor networks for telecare

K.J. Turner
University of Stirling, Stirling, Scotland, UK

4.1 Introduction

The world population is aging, with the percentage of older people gradually rising. An older population, coupled with pressure on social and health care budgets, means that care providers will be increasingly challenged to cope. As a result, it will not be feasible to provide enough senior care facilities, as they are much more expensive than providing seniors with care in their own homes. Given this trend, seniors will need more assistance to live longer in their own homes. Telecare technologies have been promoted as an important contribution to the problem of an aging population. For concreteness, illustrations are drawn from the MATCH Project: Mobilizing Advanced Technologies for Care at Home (MATCH Consortium, 2013; Turner, 2012).

Section 4.2 discusses the need for telecare technologies and how home networks support them. Section 4.3 looks at the networks used in telecare, the sensors that provide essential telecare data, the actuators that realise support for telecare, and the telecare services that can be built on them. Section 4.4 considers the systems that embed these elements. An important consideration is how telecare can be made manageable and customisable for individual needs. Section 4.5 rounds off the chapter by summarising the main ideas and identifying future trends and sources of further information.

4.2 Context of telecare

4.2.1 The aging population

The percentage of older people (over 65) is gradually rising. In the United Kingdom, for example older people comprised 24.4% in 2000, and the number is expected to rise to 39.2% by 2050 (Select Committee on Economic Affairs, 2003). In Europe, the number of older people is expected to grow from 75 million in 2004 to 133 million in 2050 (Carone & Costello, 2006). A similar situation applies in other developed countries, with much higher percentages forecast for some areas (e.g. 71.3% by 2050 in Japan). Although people are living longer, many have to deal with long-term, age-related conditions. Technology to support home care delivery can offer significant benefits.

Particularly in rural settings, the ability to support care at a distance can save the time and effort needed to travel to care facilities. Many health authorities are promoting self care at home rather than exclusively centrally provided care. In this model, trends,

Ecological Design of Smart Home Networks. http://dx.doi.org/10.1016/B978-1-78242-119-1.00004-7

anomalies, and alert conditions can be identified and reported to a central location such as a call centre or a health centre. Family members can be reassured that the user is being monitored for dangerous situations. Professional care providers can be relieved of performing low-level monitoring tasks. The seniors can therefore be supported to live longer in their own homes, where they are in familiar surroundings and near to the people and the areas they know.

Social care supports the wellbeing of individuals in the community, while health care deals mainly with the diagnosis and treatment of illness and impairment. An important part of social care is helping older people to live independently in their own homes and communities. This covers a range of factors in social and mental wellbeing including activities of daily living (ADLs) (Katz, 1983). Basic activities include eating, hygiene and mobility, while instrumental activities include communication, housework and shopping. As a person gets older, these tasks may become more difficult to achieve or maintain due to physical or mental deterioration.

4.2.2 Telecare

Technology for telecare has been enthusiastically embraced as part of the solution for the aging population. *Telecare* refers to automated support of social care at home. This includes monitoring for dangerous situations (e.g. falls, household floods and night wandering) as well as a range of services for people with physical impairment or mobility issues (e.g. curtain openers, door entry phones and home automation). Telehealth (also called *telemedicine* or *e-health*) refers to remote support of health care at home. This includes remote consultation and diagnosis (typically via videoconference) as well as monitoring health parameters and vital signs (e.g. blood pressure, heart rate and seizure risk). These data can be sent via the Internet to a general practitioner, nurse or clinic.

Although 'telecare' is the generic term used in this work, other terminology is used in this field. Technologies for assisted (or independent) living provide solutions that help older people to prolong an independent life at home. Assistive technologies are more general, comprising a variety of devices that help with daily living, such as wheelchairs and stair lifts. Home automation aims to improve management of the home (e.g. appliance control, entertainment and heating), while building control refers to the office environment. Smart homes provide a degree of intelligence and programmability. Ambient intelligence refers to the automated capabilities of a smart home.

A telecare system aims to help users to live independently and normally at home, supporting their care and wellbeing. For example, nonintrusive sensing can confirm that the individual is complying with medical advice, is sleeping well, and is dealing with personal hygiene and toileting. The system can also check for potentially hazardous situations such as a gas cooker not being lit, water being left running or the user falling. More advanced systems can help with ADLs through speech-based or visual prompting, and by providing reminders for medication, appointments, etc. Trends and anomalies can be noticed in user behaviour and reported to care providers for further investigation.

Telecare can be supported by purpose-designed computer systems (Turner & Maternaghan, 2012). Typically, some kind of home platform is provided to collect, analyse, react to and forward telecare data collected from a variety of sensors or other data sources. Besides sensor inputs, a telecare system can receive data from software services (e.g. communications media, speech recognition or weather forecasts). A telecare system is able to respond through a variety of actuators to control appliances, maintain the home environment, signal alert conditions, etc. Services can also perform output actions. More sophisticated systems have a degree of programmability, allowing customisation for individual user needs and adaptation to changing circumstances.

4.2.3   Telecare networks

A telecare system collects behavioural and environmental data from sensors within the home. This information is stored locally, for possible analysis or summarisation prior to uploading to a care centre. Telecare services act on the data to ensure the user's safety, provide reminders, look after the home, etc. The context of a sample telecare system is shown in Figure 4.1. There are two 'ecological' networks here: one within the home, and the other comprising its connection with the outside world.

Within the home, a central telecare system is connected to a variety of networks. Wireless networks might report activity around the home and collect medical data. Wired

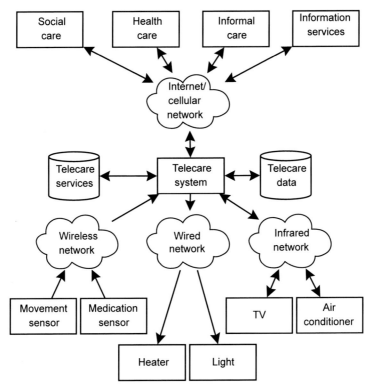

Figure 4.1 A sample telecare system in context.

networks might control household appliances. Infrared networks might communicate with audiovisual systems. Local services can manage and support care in the home.

Telecare systems connect to the outside world using an Internet connection (phone, broadband) or a cellular network. Telecare data can be presented locally (for the user or care provider) and can be sent to a social work centre. Key telecare data that could have health implications (e.g. diet, medication or sleeping) can be sent to a health centre. With the user's agreement, alerts and high-level information can also be sent to informal caregivers (e.g. family and neighbours). External services can also send useful information to the home (e.g. community information and weather forecasts).

4.2.4   The ecology of telecare networks

In a general sense, ecology deals with the relationship of organisms to their environment. In the specific context of this chapter, the *ecology of a telecare network* refers to how people relate to the care services provided within the home and externally, and how cared-for people relate to those who provide informal and formal care.

A telecare network mediates these relationships through the connections between the telecare system, home devices and services and external care services. The ecology of a telecare network thus refers to the relationship among the devices and services used to support telecare, and their context in the wider world of care and society.

A number of environmental goals can be supported through telecare. Particularly in a rural environment, providing home care can require substantial travel. The end user might have to travel to distant clinics for health care, and social workers may have to travel to remote areas. Telecare allows certain aspects of care to be remotely configured and monitored. Visits to the home can be reduced, since a simple reconfiguration (e.g. medication reminder times) can be undertaken remotely. As an example of automated routine monitoring, a user who is prone to bed-wetting or falls does not need unnecessary visits to check if this needs attention; only if a sensor reports such an event is a visit needed. More importantly, an automated alert allows a speedier response.

A telecare system can also manage energy use in the home more efficiently. The house can be heated only as required, being turned on only as needed (e.g. if frost is forecast). Conversely, the house can be cooled by opening windows instead of using energy-hungry air conditioning. Indeed, the user can choose the balance between energy-saving measures and other goals that may affect energy usage.

Telecare also supports people in relating to their communities. For example, telecare can support community TV, local news and helping community members to stay in touch. The technological capabilities of telecare can support 'care in the community,' the collective responsibility for providing care within a neighbourhood.

4.3   Telecare networks

4.3.1   Home networks for telecare

Many kinds of home networks are used in telecare, typical examples being given below. The choice of network is usually dictated by what a manufacturer's sensors and actuators use, although there are some common standards. In practice, multiple networks often have to be deployed in the home and integrated through the telecare system. Wired solutions are the obvious choice for new-build housing or where standard household wiring can be used. In existing houses, wireless is preferable to minimise the disruption of new wiring. However, battery life can be an issue with wireless sensors; typically this is 1–3 years, and requires an on-site visit to replace old batteries.

Wired Networks: At the least, the home will have phone and electrical mains wiring that can be used for digital communication. The internal phone network will typically provide Internet access via a Digital Subscriber Line (DSL) (Wikipedia, 2013a). Mains wiring can support communication using standards such as HomePlug (HomePlug Alliance, 2013), and can also be used to control mains appliances with standards like X10 (Wikipedia, 2013b). Particularly in new buildings, it makes sense to install structured cabling during construction (Lalena, 2013); Cat 5e and Cat 6 cabling are common. Ethernet, more correctly Carrier Sense Multiple Access with Collision Detection (CMSA/CD), is widely used for computer communication in the home (Institution of Electrical and Electronics Engineers, 1984). Universal Serial Bus (USB) is also widespread, but is normally used for only short-range communication (USB Implementers Forum, 2013). It is, however, possible to use extenders to achieve USB communication over a longer range.

Wireless Networks: In a domestic setting, wireless networks are used for short-range communication (10–100 m). Wireless networks common in telecare include Bluetooth (Bluetooth Special Interest Group, 2013), Wi-Fi (Institution of Electrical and Electronics Engineers, 2012) and ZigBee (ZigBee Alliance, 2013). For domestic wireless sensors, 433 MHz and 866–868 MHz are widely used from the Industrial, Scientific and Medical (ISM) bands. Many manufacturers supply proprietary equipment operating at these frequencies (e.g. Oregon Scientific, 2013; Visonic, 2013). USB can also be carried over short-range, wideband, wireless networks. Cellular networks are common for collecting and distributing information within and outside the home.

Infrared Networks: Infrared is normally used point to point, typically to control domestic appliances such as audiovisual equipment or air conditioners. Signaling protocols vary quite widely, but are governed by a set of industry standards (Infrared Data Association, 2013).

Home Automation Networks: These are designed for controlling the home or other buildings. As an example, KNX (Konnex Association, 2013) derives from earlier work on the European Installation Bus. It supports a variety of media and devices, being widely used in building management including domestic applications. Other examples of such networks include, for example, Lonworks (Echelon Corporation, 2013), a set of networking standards for applications such as building management, home automation and transportation.

4.3.2 Sensors for telecare

Telecare uses a mixture of sensors designed specifically for telecare/telehealth and sensors from applications such as home automation, home security and environmental monitoring. The following examples are typical of what might be found in a telecare setting.

Location: A user's location can be sensed using pressure mats (e.g. on the floor, a chair or a bed). Force plates similarly detect that a household item has been used (e.g. a kettle was picked up). Radio frequency identification (RFID) tags and smart badges can also be used, whether with active or passive transceivers. However, the user must remember to wear the tag or badge, which can be problematic with a forgetful older person. As a less intrusive solution, face and skeleton tracking are possible with a camera. Video processing with a standard camera can be challenging, but a device such as the Kinect (Microsoft, 2013a) makes things easier, as it has access to depth information. Indeed, the Kinect can recognise up to six people within view and track two of them. The ability to recognise people is valuable if the cared-for person lives with others (a common challenge for many telecare systems).

Movement: Movement can be detected by simple passive infrared sensors (PIRs). The tracking offered by the Kinect is a more sophisticated way of detecting the movement of individuals. The Kinect can also be used to recognise gestures, for example, to control devices around the home (Haag, 2012). The Wii Remote (Nintendo, 2013) likewise offers opportunities for gesture recognition.

Safety and Security: Standard safety and security devices are also useful in telecare. Apart from PIRs, sensors are commonly deployed in the home to detect gas leaks, smoke and flooding (e.g. Visonic, 2013). Magnetic reed switches can be fitted to doors and windows, aiming to improve security by warning that something has been left open.

Appliances: When used with cupboards or domestic appliances, switches are also useful for detecting activities around the home (e.g. using the refrigerator, opening a pantry door or flushing a toilet). Measuring appliance power consumption, for example, with a Plugwise module (Plugwise, 2013), is another way of recording the user's activities around the home. For example, this device can detect use of a mains-powered telephone, a microwave oven or a TV. An infrared transceiver such as the IRTrans (2013) can monitor signals sent by infrared remote controls. This allows, for example, use of a TV to be observed so that the telecare system determines what programs the user likes to watch.

Environment: Digital temperature and humidity sensors (thermohygrometers) provide information about conditions around the house (e.g. Oregon Scientific, 2013). A central heating system or a climate control system will, of course, have its own built-in sensors. However, separate sensors are able to provide a finer-grained picture of conditions in the house. This could be helpful for an older person who is forgetful about turning the heat on in colder weather. Light sensors can be used to maintain lighting levels, or simply to distinguish daylight from darkness. Sound sensors can be relevant, for example, to determine that windows should be shut to reduce outside noise. Although there may not currently be specialised sound monitors dedicated for telecare, a microphone can be used with standard computer audio facilities.

Communication: Standard communication services are used for telecare, for example, e-mail, short messaging service (SMS) and audio/video conferencing. Like community TV, these allow the user to stay in touch with family, friends and the community.

Speech recognition allows use of verbal commands, which may be a necessity if the user has physical disabilities. Voice recognition (i.e. identifying the speaker) may also be useful. A variety of 'Internet buddies' can be used for communication by people who are not computer literate. For example, the Nabaztag 'rabbit' and its successor the Karotz (Violet, 2013) offer a variety of sensors: buttons, speech input and RFID. The Tux Droid 'penguin' is a user-friendly sensor device that can also be used in this capacity.

Health: A variety of medical sensors are used to monitor vital signs or medical conditions like heart rate, blood sugar and lung capacity. Other sensors detect conditions such as enuresis (bed-wetting), epileptic seizure and heart problems. Automated medicine dispensers can sense that the user has taken medication, leading to a reminder if the user forgets.

4.3.3   Actuators for telecare

It is common to talk of sensor networks, and indeed sensors predominate in the home. However, it is desirable to have actuators as well, to respond to the sensed input. The following describes a range of actuators that might be used in telecare.

Appliances: In general, household appliances can be controlled via infrared (IRTrans, 2013), Plugwise modules (Plugwise, 2013) or X10 (Wikipedia, 2013b). For example, a forgetful user might appreciate having favourite programs recorded automatically, or the coffee percolator might be switched on when the user gets home. Curtain and blind openers are helpful for someone with limited mobility. Door entry phones coupled with remotely controlled locks are similarly useful; these also allow doors to be automatically locked if the house is left unoccupied.

Environment: Heating, ventilation and air conditioning can be automatically controlled via infrared, Plugwise modules or X10 to suit the user. Lighting levels can also be adjusted via X10.

Communication: Bidirectional communication services such as e-mail and SMS can also deal with output. Multimodal (multisensory) interfaces can help someone with a disability, or simply with particular communication preferences. Speech synthesis and chimes can be used for audio output. Mobile devices or specialised ones like the SHAKE (Sensing Hardware Accessory for Kinaesthetic Expression) can be used for tactile output, for example, a vibration as a reminder. Scent dispensers (e.g. Dale Air, 2013) can be used for olfactory output: an appetising aroma might remind a user of mealtime. An Internet buddy, mentioned earlier, can also be used for output. The i-Buddy 'angel' (Union Creations, 2013) can provide simple signals, for example, to indicate message arrival. Other devices like the Nabaztag or Karotz 'rabbit' and the Tux Droid 'penguin' have more sophisticated possibilities, such as monitoring of gestures and spoken output.

4.3.4   Services for telecare

External services can provide data much like sensors. For example, a weather forecast service might be used to adjust the heating, or a travel information service might alert the user to leave earlier for an appointment due to traffic delays. External services can also consume data much like actuators. As examples, a health alert service might act on uploaded vital signs data and a shopping service might act on information about what supplies are running low.

Telecare services can run locally on the telecare system. A night wandering service, for example, can deal with a common problem for older people: getting up in the middle of the night and thinking that it is time to go out. In such a case, a family member's voice might tell the user that it is not yet time to get up. A reminder service might help the user to keep appointments or to take medicine on time. A safety service might check for gas or flooding, or might advise the user that a window has been left open on leaving the house. Telecare services can be also be created hierarchically, building on lower-level services. Thus, a movement sensing service might be combined with a diary service to report an intruder while the house is unoccupied.

Services can be defined to transform multiple low-level inputs into high-level inputs. As an example, a fall service might combine information from several sources: a fall detector, a movement detector and a video camera. This low-level data can be combined to give a reliable indication that the user has fallen. As a further example, a house occupancy service might use information from movement, door and appliance sensors to decide whether the house is unoccupied. Combining multiple sensor inputs is called *sensor fusion*. The converse term, actuator fusion, refers to turning a high-level output into multiple low-level outputs (Turner, 2010). Thus, a high-level action such as 'remind the user of a 2:00 pm nurse appointment' might be achieved through a visual or spoken output if the user is around the house, or through a text message if the user is away from home.

4.4   Telecare systems

4.4.1   Telecare platforms

Unsurprisingly, there is no industry-standard platform for telecare: The need is for interoperability rather than uniform design. Individual companies therefore develop their own proprietary solutions (e.g. Cisco, 2013; Docobo, 2013; Intel, 2013; Microsoft, 2013b; OmniQare, 2013). However, OSGi (originally Open Systems Gateway initiative) has proven popular for telecare in many research projects, and also in some commercial solutions.

OSGi (OSGi Alliance, 2013) was developed as a service-oriented, dynamic module system for Java. OSGi modules are called bundles, and have a simple interface. Bundles can communicate via an event bus using the Event Admin service. This allows bundles to be created in a loosely coupled way, and yet to cooperate in support of higher-level services. OSGi was originally designed for services in the home, but has been adopted for other applications such as services in vehicles.

The basic bundle interface needs to be augmented for easy configuration and integration. SODA (Service-Oriented Device Architecture) provides a framework for uniform development of OSGi device bundles (de Deugd, Carroll, Kelly, Millett, &

Ricker, 2005); this has been enhanced with data semantics (Gouvas, Bouras, & Mentzas, 2007). Other work has aimed to make bundles self-describing, thereby easing their management and allowing arbitrary bundles to be controlled by a rule-based system (Maternaghan & Turner, 2011; McBryan & Gray, 2012).

Besides home automation, OSGi has also been used to support telecare and telehealth. ATLAS (Kind, Bose, Yang, Pickles, & Helal, 2006) is a well-known example that uses OSGi. This effort started as an academic project to support sensor-actuator networks in a service-oriented manner, but has been made available commercially. Among other applications, the ATLAS middleware has been used for smart homes and healthcare. Other projects that have used OSGi include e-HealthCare (2013), MATCH (Gray et al., 2007) and SAPHIRE (Hein et al., 2006).

Context-aware systems aim to make a system reactive to context, and have been used in home applications. Gaia (Román et al., 2001) supports 'active spaces' that rely heavily on contextual information, including presence. EasyLiving (Brumitt, Meyers, Krumm, Kern, & Shafer, 2000) is designed to support intelligent environments through dynamic interconnection of a variety of devices. This middleware offers mechanisms such as intersystem communication, location tracking for objects and people and visual perception.

4.4.2 Managing telecare

Telecare systems are proprietary and often require specialised technical expertise to modify. As a result, it may not be easy to customise the system for individual care needs. If the user has a long-term and perhaps degenerative condition, the system should adapt to needs that evolve over time. One solution that has been proposed is the use of policies as user-definable rules for care.

Policies are computer-interpreted rules that are automatically executed when events occur. Policies have been used in applications such as access control, network/system management and quality of service. As a typical example, Ponder (Damianou, Lupu, & Sloman, 2001) is a general-purpose policy approach. It offers a mature methodology for policies in applications such as system management and sensor networks. However, nearly all policy approaches are designed for technical applications. A different kind of policy approach is needed for the 'softer' management tasks found in human-oriented applications such as home care.

A few systems employ rules for managing the home. A rule-based system (Leong, Ramli, & Perumal, 2009) was described for smart homes. However, this is a rather heavyweight solution that expects home devices to be interconnected via an Ethernet. Although the system supports basic rules, these do not seem to be defined by end users. Gadgetware (Kameas, Mavrommati, & Markopoulos, 2005) achieves a similar result, although not in a recognisably rule-based way. Physical objects are given a digital representation with 'plugs' that can be connected via 'synapses.'

As a concrete example of managing telecare, Advanced Component Control Enhancing Network Technologies (ACCENT) is an approach and a set of tools for managing systems through goals and policies (Turner, 2013a). ACCENT and its accompanying policy language, Adaptable and Programmable Policy Environment and Language (APPEL), have been used to support telecare (Turner, 2013b).

Figure 4.2 shows the high-level architecture of the ACCENT system. When used to support care, this operates in a home environment: the user's home and its wider context, including times when the user is not at home. Sensors report medication dispensing, appliance usage, activities around the home, etc. Actuators send alerts, provide reminders, manage appliances, etc. Services supplement these with additional inputs and outputs. The policy server achieves high-level user goals by executing appropriate lower-level policies. The event transformer provides services that map between low-level signals and high-level ones.

All ACCENT components are implemented as OSGi bundles and linked via the OSGi event bus. This conveys inputs and outputs among all components, but particularly to/from the policy server. Usually, the policy server receives inputs from sensors (and services) that trigger one or more goals and policies. These determine how the system should react to the input. The resulting output actions are then performed by the relevant actuators (and services).

The event transformer achieves sensor fusion by intercepting one or more signals and transforming them into high-level inputs for the policy server. Conversely, the event transformer realises actuator fusion by taking high-level outputs from the policy server and converting them into several low-level signals. Sensors, actuators and services normally communicate only with the policy server or event transformer. However, they can also communicate directly with each other to build higher-level capabilities.

Goals and policies are the primary way for users to manage the ACCENT telecare system. Among other things, the APPEL language is used to define goals and policies of various kinds (Turner, 2013b). Although these can be created from scratch, a library has been developed with over 100 predefined templates for ease of use. The user generally just needs to select a template, defining key values such as an emergency telephone number or the user's normal bedtime.

Goals are high-level user objectives. An optional condition specifies the circumstances in which the goal applies. Only two types of actions are used: to maximise or to minimise some measure of a goal. These measures are defined in terms of factors that affect goal achievement. Most approaches to goals are based on logic, but ACCENT

Figure 4.2 High-level ACCENT
system architecture (Turner, 2010).

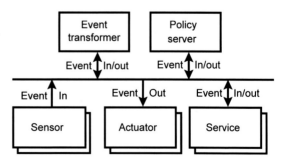

treats goal achievement as an optimisation problem. This allows goals to be realised in a dynamic way depending on current circumstances (which may vary over time). It is also pragmatic in that goals are achieved as far as is possible, and do not need to be completely fulfilled in some absolute sense.

As an example, suppose the user has the goal of staying comfortable. This aim is high level and not directly executable. To give the goal meaning, the user has to specify the factors that contribute to the goal (i.e. to its measure). For example, the user may state that being comfortable depends on the indoor temperature (e.g. with ideal value 21 °C), the audio volume (e.g. noise significant only above a threshold of 70 dB) and the risk of getting a chill outdoors. Each of these factors is internally associated with a weight that is automatically determined by the system so that the factors make similar contributions.

A user can define multiple goals, each of which has a relative importance assigned by the user. For example, the user may decide that staying comfortable is twice as important as saving energy. The weighted combination of goal measures constitutes an overall evaluation function to be optimised dynamically by the system.

Goals are achieved by special policies called *prototypes*. Like all policies, prototypes normally have a trigger, a condition and an action (each of which may be composite). Policies can also be associated with profiles such as 'at home' or 'on holiday.' Unlike other kinds of policies, prototypes indicate how they affect goals through their effect on the factors in goal measures.

As an example, a prototype might be defined to ensure that the house does not become overheated. If the indoor temperature is reported as hot, the air conditioning can be set to high. This might have the effects of reducing the temperature by 4 °C and increasing energy consumption by 4 kWh (which are factors in goals such as staying comfortable or saving energy).

A regular policy is similar to a prototype but does not identify an effect on goals. This is used for policies that should always apply irrespective of the current goals. As an example, if an older person needs to go to the toilet at night, then there is a risk that the person will fall in the darkness. A policy can be defined to monitor a bed occupancy sensor. If this reports that the user has gotten out of bed at night, the toilet light can be switched on. When the bed becomes occupied again, the toilet light can be switched off.

When triggers occur, they cause regular policies and goal-related (prototype) policies to be activated. Prototypes are selected to maximise the overall evaluation function, and are then combined with regular policies. This allows the system to react appropriately and dynamically to changing circumstances.

Users interact with the policy system via a number of alternative interfaces. Several wizards ease the task of defining and editing the goals and policies. The web wizard is the most comprehensive: it is multilingual, and can be used remotely (perhaps by a care worker). As an example of this wizard, Figure 4.3 shows the goals currently defined by a user. Goals can be edited by clicking on their labels, disabled and deleted. The relative importance of each goal can be set by dragging a slider.

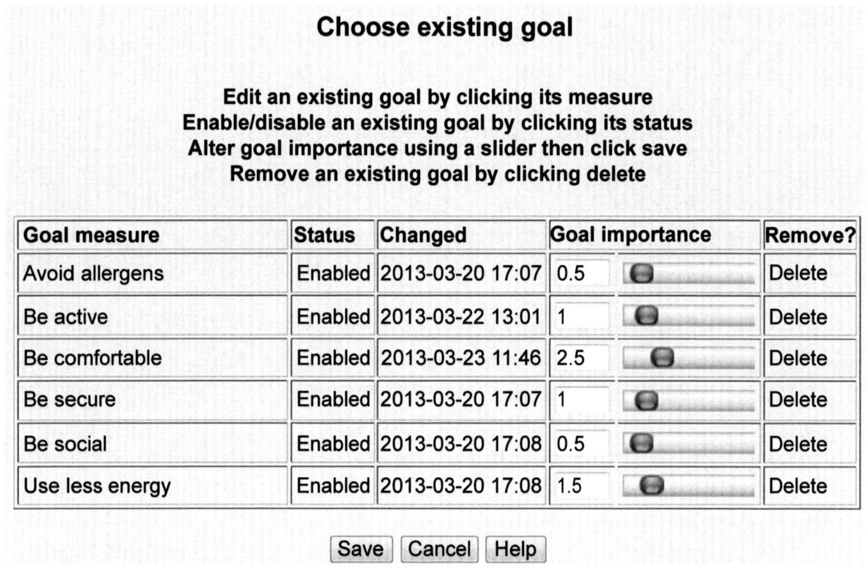

Choose existing goal

Edit an existing goal by clicking its measure
Enable/disable an existing goal by clicking its status
Alter goal importance using a slider then click save
Remove an existing goal by clicking delete

Goal measure	Status	Changed	Goal importance		Remove?
Avoid allergens	Enabled	2013-03-20 17:07	0.5		Delete
Be active	Enabled	2013-03-22 13:01	1		Delete
Be comfortable	Enabled	2013-03-23 11:46	2.5		Delete
Be secure	Enabled	2013-03-20 17:07	1		Delete
Be social	Enabled	2013-03-20 17:08	0.5		Delete
Use less energy	Enabled	2013-03-20 17:08	1.5		Delete

Save | Cancel | Help

Figure 4.3 Web wizard: editing a list of telecare goals.

4.5 Conclusion

4.5.1 Summary

The world population is aging, with the percentage of older people gradually rising. The growing number of seniors, coupled with pressure on social and health care budgets, means that care providers will be increasingly challenged to cope. Telecare technologies have been promoted as an important part of the solution.

The chapter has discussed the ecology of home sensor networks for telecare, that is, the devices and services used to support telecare, their relationships and their context in the wider world of care and society. Telecare has been described as a computer-supported approach that remotely delivers social care to the home. A telecare system links a variety of devices and services in the home, and interacts with the outside world: (in)formal care and external services. The types of networks used in telecare have been introduced. The kinds of sensors, actuators and services deployed in telecare have also been described.

Although there is no industry standard for telecare platforms, OSGi has been introduced as a popular basis for research (and some commercial) solutions. Components in a telecare system are then OSGi bundles that can communicate via an event bus. The approach of the ACCENT system, in particular, has been described. This system can be customised and adapted through user-defined goals and policies.

4.5.2 Future trends

Although significant progress has been made in sensor networks for telecare, a number of pragmatic and research challenges still need to be addressed. Barriers to telecare uptake have also been identified (Clark & McGee-Lennon, 2011; Taylor & Yadav, 2013).

Usability and user acceptance are important issues. User-centred design and codesign should be adopted to ensure that telecare devices and services are acceptable to end users. It will also be vital to create solutions that are reusable and do not meet just a single need. An important development will be in creating platforms that make it easier to combine telecare solutions from multiple suppliers.

Personalisation will require further research into how best to customise telecare systems for individual and evolving needs. Policy-based management and context awareness have both shown promise for this. Telecare systems should become able to learn, through observation, what the user's needs and preferences are.

Multimodal interfaces are likely to be used increasingly in telecare, and in fact in all computing. Speech input–output and audio in general are already widely available. Gesture recognition has become common in games systems. Other modalities such as tactile or olfactory output are also likely to find uses in telecare.

Standards for telecare systems are at an early stage. As a result, interoperability and data interchange are significant problems. Progress is being made in standards for telehealth (Continua Health Alliance, 2013), but corresponding standards for telecare still need to be defined. In health care, there are standards for data interchange such as Health Level 7 (American National Standards Institute, 1999). Again, telecare will need equivalent standards. This is particularly important because health care and social care in many countries have traditionally been separate and do not share data.

Many care professionals were trained during a time when telecare barely existed. There is, therefore, a need for courses that retrain these professional as well as educate the new generation of telecare practitioners. The evidence base for telecare is gradually being built up; however, this effort needs to be enriched with solid evidence for when and how to use telecare, along with convincing cost–benefit analyses. Public awareness of telecare is also desirable so that realistic expectations can drive future developments.

Ethics and privacy are sensitive issues that will require further work on guidelines for telecare. Some users have expressed concerns that telecare is a form of 'state surveillance' (Sorell & Draper, 2012). Standards also need to be defined for using telecare data in an appropriate manner. For example, a social worker might need to be aware of health data that would normally be considered confidential.

4.5.3 Further information

Besides specific citations throughout the chapter, the following more general sources will amplify the main points. Akyildiz and Vuran (2010) discuss wireless sensor networks in general. Oudshoorn (2011) provides insight into how telecare can transform care delivery. Taylor and Yadav (2013) provide a useful review of how telecare is becoming mainstream. Turner (2012) covers advances in home care technologies, particularly for telecare. Yacoub and Yang (2006) illustrate the increasing trend toward use of body sensor networks to monitor health and wellbeing.

References

Akyildiz, I. F., & Vuran, M. C. (2010). *Wireless sensor networks.* Chichester, UK: John Wiley and Sons.

American National Standards Institute. (1999). *Health level seven—An application protocol for electronic data exchange in healthcare environments.* Washington DC, USA: American National Standards Institute.

Bluetooth Special Interest Group. (2013). *Bluetooth technology.* http://www.bluetooth.com. Accessed August 2013.

Brumitt, B., Meyers, B., Krumm, J., Kern, A., & Shafer, S. (2000). Easyliving: technologies for intelligent environments. In P. J. Thomas & H.-W. Gellersen (Eds.), *Proc. 4th Int. symp. on handheld and ubiquitous computing, pp. 12–29, number 1927 in lecture notes in Computer Science.* Berlin, Germany: Springer.

Carone, G., & Costello, D. (2006). Can Europe afford to grow old? *Finance and Development Magazine, 43*(3), 8–31.

Cisco. (2013). *HealthPresence.* http://www.cisco.com/web/strategy/healthcare/cisco_health presence_solution.html. Accessed August 2013.

Clark, J. S., & McGee-Lennon, M. R. (2011). A stakeholder-centred exploration of the current barriers to the uptake of home care technology in the UK. *Assistive Technologies, 5*(1), 12–25.

Continua Health Alliance. (2013). *Connected health standards.* http://www.continuaalliance.org. Accessed August 2013.

Dale Air. (2013). *Dale air dispenser.* http://www.daleair.com. Accessed August 2013.

Damianou, N., Lupu, E. C., & Sloman, M. (2001). The Ponder policy specification language. In *Proc. policy workshop, number 1995 in lecture notes in Computer Science.* Berlin, Germany: Springer.

de Deugd, S., Carroll, R., Kelly, K., Millett, B., & Ricker, J. (2005). Soda: service oriented device architecture. *Pervasive Computing, 5*(3), 94–96.

Docobo. (2013). *Telehealth solutions.* http://www.docobo.com. Accessed August 2013.

e-HealthCare. (2013). *e-HealthCare project.* http://ehealth.sourceforge.net. Accessed August 2013.

Echelon Corporation. (2013). *Lonworks.* http://www.echelon.com/applications/smart-control. Accessed August 2013.

Gouvas, P., Bouras, T., & Mentzas, G. (2007). An OSGi-based semantic service-oriented device architecture. In R. Meersman, Z. Tari, & P. Herrero (Eds.), *On the move to meaningful Internet systems, pp. 773–782, Number 4806 in lecture notes in Computer Science.* Berlin, Germany: Springer.

Gray, P. D., McBryan, A., Hine, N. J., Martin, C. J., Gil, N., Wolters, M., et al. (2007). *A scalable home care system infrastructure supporting domiciliary care.* Technical Report CSM-173. UK: Computing Science and Mathematics, University of Stirling.

Haag, T. (2012). Kinect in home automation (BSc Dissertation). Computing Science and Mathematics, University of Stirling, UK.

Hein, A., Nee, O., Willemsen, D., Scheffold, T., Dogac, A., & Laleci, G. B. (2006). Intelligent healthcare monitoring based on semantic interoperability platform—the homecare scenario. In *Proc. 1st European conf. on eHealth, Fribourg, Switzerland.*

HomePlug Alliance. (2013). *HomePlug technology.* https://www.homeplug.org. Accessed August 2013.

Infrared Data Association. (2013). *Infrared technology.* http://www.irda.org. Accessed August 2013.

Institution of Electrical and Electronics Engineers. (1984). *Local and metropolitan area network standards—Carrier sense multiple access with collision detection (CSMA/CD) access method and physical layer specifications, IEEE 802.3*. New York, USA: IEEE Press.

Institution of Electrical and Electronics Engineers. (2012). *Wireless LAN medium access control and physical layer specifications, IEEE 802.11*. New York, USA: IEEE Press.

Intel. (2013). *Home health reference platform*. http://www.intel.com/content/www/us/en/intelligent-systems/medical-applications/home-health-reference-platform-featuring-realtek-rtl8954c-ioh.html. Accessed August 2013.

IRTrans. (2013). *IRTrans devices*. http://www.irtrans.com. Accessed August 2013.

Kameas, A., Mavrommati, I., & Markopoulos, P. (2005). Computing in tangible: using artifacts as components of ambient intelligence environments. In G. Riva, F. Vatalaro, F. Davide, & M. Alcañiz (Eds.), *Ambient intelligence: The evolution of technology, communication and cognition* (pp. 121–142). Amsterdam, Netherlands: IOS Press.

Katz, S. (1983). Assessing self-maintenance: activities of daily living, mobility, and instrumental activities of daily living. *Journal of the American Geriatrics Society, 31*(12), 721–727.

Kind, J., Bose, R., Yang, H.-I., Pickles, S., & Helal, A. (2006). Atlas: a service-oriented sensor platform. In *Proc. Workshop on practical issues in building sensor network applications*. Los Alamitos, California, USA: IEEE Computer Society.

Konnex Association. (2013). *KNX standard for home and building control*. http://www.knx.org. Accessed August 2013.

Lalena. (2013). *Structured wiring*. http://www.structuredhomewiring.com. Accessed August 2013.

Leong, C., Ramli, A. R., & Perumal, T. (2009). A rule-based framework for heterogeneous subsystems management in smart home environment. *IEEE Transactions on Consumer Electronics, 55*(3), 1208–1213.

MATCH Consortium. (2013). *MATCH (Mobilising advanced technologies for care at home)*. http://www.match-project.org.uk. Accessed August 2013.

Maternaghan, C., & Turner, K. J. (2011). Pervasive computing for home automation and telecare. In S. I. A. Shah, M. Ilyas, & H. T. Mouftah (Eds.), *Pervasive communications handbook* (pp. 17.1–17.25). Boca Raton, FL, USA: CRC Press.

McBryan, A., & Gray, P. D. (2012). Dynamic configuration of home services. In K. J. Turner (Ed.), *Advances in home care technologies: Results of the MATCH project* (pp. 86–105). Amsterdam, Netherlands: IOS Press.

Microsoft. (2013a). *Kinect*. http://www.microsoft.com/en-us/kinectforwindows. Accessed August 2013.

Microsoft. (2013b). *HealthVault*. https://www.healthvault.com. Accessed August 2013.

Nintendo. (2013). *Wii remote*. http://www.nintendo.com. Accessed August 2013.

OmniQare. (2013). *iQare system*. http://www.omniqare.com. Accessed August 2013.

Oregon Scientific. (2013). *Wireless technology*. http://www.oregonscientific.com. Accessed August 2013.

OSGi Alliance. (2013). *OSGi system*. http://www.osgi.org. Accessed August 2013.

Oudshoorn, N. (2011). *Telecare technologies and the transformation of healthcare*. Basingstoke, UK: Palgrave Macmillan.

Plugwise. (2013). *Plugwise technology*. http://www.plugwise.com. Accessed August 2013.

Román, M., Hess, C. K., Cerqueira, R., Ranganathan, A., Campbell, R., & Nahrstedt, K. (2001). Gaia: a middleware infrastructure for active spaces. *Pervasive Computing, 1*(4), 74–83.

Select Committee on Economic Affairs. (2003). *Aspects of the economics of an ageing population*. London, UK: Stationery Office Limited.

Sorell, T., & Draper, H. (2012). Telecare, surveillance, and the welfare state. *American Journal of Bioethics*, *12*(9), 36–44.

Taylor, K., & Yadav, A. (2013). *Telecare and telehealth—A game changer for health and social care*. London, UK: Deloitte Centre for Health Solution.

Turner, K. J. (2010). Device services for the home. In K. Drira, A. H. Kacem, & M. Jmaiel (Eds.), *Proc. 10th Int. conf. on new technologies for distributed systems* (pp. 41–48). Los Alamitos, CA, USA: IEEE Computer Society.

Turner, K. J. (Ed.). (2012). *Advances in Home Care Technologies: Results of the MATCH project*. Amsterdam, Netherlands: IOS Press.

Turner, K. J. (2013a). *ACCENT (Advanced component control enhancing network technologies)*. http://www.cs.stir.ac.uk/accent. Accessed August 2013.

Turner, K. J. (2013b). *APPEL (Adaptable and programmable policy environment and language)*. http://www.cs.stir.ac.uk/appel. Accessed August 2013.

Turner, K. J., & Maternaghan, C. (2012). Home care systems. In K. J. Turner (Ed.), *Advances in Home Care Technologies: Results of the MATCH project* (pp. 21–29). Amsterdam, Netherlands: IOS Press.

Union Creations. (2013). *i-Buddy*. http://www.i-buddy.com. Accessed August 2013.

USB Implementers Forum. (2013). *Universal serial bus*. http://www.usb.org. Accessed August 2013.

Violet. (2013). *Karotz rabbit*. http://store.karotz.com/en_GB. Accessed August 2013.

Visonic. (2013). *Wireless technology*. http://www.visonic.com. Accessed August 2013.

Wikipedia. (2013a). *Digital subscriber line*. http://en.wikipedia.org/wiki/Digital_subscriber_line. Accessed August 2013.

Wikipedia. (2013b). *X10 industry standard*. http://en.wikipedia.org/wiki/X10_(industry_standard). Accessed August 2013.

Yacoub, M., & Yang, G.-Z. (2006). *Body sensor networks*. Berlin, Germany: Springer.

ZigBee Alliance. (2013). *ZigBee technology*. http://www.zigbee.org. Accessed August 2013.

Smart home networking for content management

E.A. Heredia
Advanced Technology Lab, Samsung Research America, Mountain View, CA, USA

5.1 Introduction

One interesting aspect of modern life is our expectation that, at home, we need to have a variety of good and efficient methods of entertainment. At home, users engage in multiple entertaining activities like watching television (TV), listening to music, viewing content on the Internet, interacting with mobile apps, participating in social networks, and playing electronic games in phones or consoles.

In the past, each of these activities had to be performed on dedicated devices, but today we are quickly moving into a more integrated environment, where users can watch a video, listen to audio, read an electronic book, or play a game originating from a number of sources and rendered on any of a number of target devices (phones, tablets, TVs, computers, game consoles, etc.).

The ecosystem of devices that we use at home for entertainment is changing quickly. There are at least four major forces acting simultaneously to change this ecosystem: connectivity, fusion, intelligence, and blending.

Connectivity is the term we use to identify the fact that many of our home and mobile devices now connect to the network, and can exchange data across the network. Although most devices connect with central home routers, typically using Wi-Fi, some devices additionally connect with others in peer-to-peer modes. Connectivity is still evolving, and there is a need for very clever solutions that will make the network much more reliable and simpler to handle. Today, it is still difficult for a majority of users to properly manage and maintain the connectivity aspect of our home networks.

Fusion is the term we use to identify the multiservice capabilities of modern devices. Tablets can serve as TVs, TVs can be used to engage in videoconferences, and phones can serve as remote controls for gaming consoles. Users now expect their devices to provide a multiplicity of services, with many of the services becoming available through downloadable applications.

Intelligence is the term we use to identify the fact that devices become more able to manage our habitat automatically. The increased computational power available in home and portable devices allows us to design much more sophisticated software that will change the way in which people interact with devices and networks.

Blending is the term we use to identify the need for technology to be hidden in a user's habitat. Modern entertainment devices and networks need to blend with our habitat so as to be inconspicuous. Some devices will blend with our common wearable objects; others will blend with the rooms and walls in our buildings. Input systems like

Ecological Design of Smart Home Networks. http://dx.doi.org/10.1016/B978-1-78242-119-1.00005-9

mice and remote control will be replaced by voice commands or intent recognition systems. In the future, any of our walls may serve as TVs. Computing devices, as well as storage, will be everywhere hidden from our view but ready to serve our needs.

In this document, we describe progress and research directions in several of these areas. In a home with connected devices, how do we design software platforms to manage our content? How do we access content services, and how do we identify new network functionality? In the past 10 years there has been significant progress to understand and address the need of media-centered home networks. Despite all the progress, most practitioners agree that a solution that comprehensively satisfies all needs has not yet been reached. There are still multiple technical and business hurdles before a truly smart home, with content freely flowing from device to device, will become the norm for the average user.

About 10 years ago, the computer, communications, and consumer industries noticed the rapid adoption of network technologies, especially Wi-Fi, in the home environment. The industry at that time realized that there was an opportunity to create a common standard that would enable fluid media operations among connected devices. Content in computers could flow to TVs, and pictures from cameras could be transferred for storage to nearby PCs. TV programs could flow from TVs and set-top boxes to phones and tablets. The notion of a fully interconnected media-centric home motivated the creation of an organization known as the Digital Living Network Alliance (DLNA), with the goal of developing the necessary common protocols required for this interoperable infrastructure (DLNA, 2003).

The original vision of the DLNA was to replicate, to a certain level, the Internet modus operandi, where the design and implementation of a collection of standardized protocols like TCP/IP, HTTP, and HTML enable fluid operations between devices (DLNA, 2003; Heredia, 2011).

After more than 10 years of standardization and development activities, the DLNA protocols have been implemented and commercially deployed in many devices, including TVs, phones, computers, stereo equipment, speakers, and tablets. A recent DLNA report shows that, as of 2013, the total number of DLNA-enabled devices sold or provided to users exceeds 2 billion (DLNA, 2013). Although iOS devices do not support the DLNA protocols, there are multiple applications that can be used to add this functionality.

However, despite the ubiquity of DLNA-enabled devices, many users are still unable to perform media exchange operations over home networks on a daily basis. Users are genuinely interested in viewing pictures, videos, and TV programs in distributed target devices such as large-screen TVs, phones, and tablets. Users are also interested in playing music stored in phones or streamed from the Internet in distributed target devices such as networked wireless speakers, connected radios, and connected stereo systems. The DLNA architecture allows many of these scenarios, but average users are often unaware of the availability of these features on their devices – or if they are aware, there are several factors that encumber the experience.

There are several alternatives to DLNA in the market today. Apple provides AirPlay, a simple and consistent solution for sharing content between Apple's portable devices (iPhone, iPad, and the iTunes app in PCs and MACs) and a networked player known as AppleTV that users connect to their TVs via HDMI. Users can browse and select personal pictures, personal videos, and commercial video content acquired from the iTunes store. The selected content can be redirected to the TV via the AppleTV device. Apple offers a high-quality user experience, but at the cost of limiting the ecosystem participants to only Apple or Apple-approved devices. Users, on the other hand, have a strong expectation that all devices, regardless of their brands or types, will work seamlessly in the smart homes of the future.

More recently, Google has introduced Chromecast, a small inexpensive USB device that plugs into the USB port of a TV. This device acts as a networked receiver. Users operate a tablet, phone, or laptop to browse content from certain apps like YouTube or Netflix. The selected content can be redirected to the TV using the Chromecast device. As of today, the domain of applicability of this solution is limited to only certain applications. It does not offer users the option to project their own pictures or their own videos. It does not give users the option to transfer TV content from one device to another, etc.

Besides DLNA's, Apple's, and Google's there are a number of alternatives that have been deployed in the market in the last 10 years. However, despite continual improvements, the different solutions have not yet reached the minds and hearts of the average consumer. There are people who use these technologies relatively frequently, but daily usage for average users remains elusive. Content management solutions at home have not yet reached the critical mass to become truly domestic. Furthermore, the fact that the current technologies work separately and independently of each other creates a fragmented landscape that makes it more difficult for mass adoption. Existing solutions are like islands of innovation unable to connect with each other and unable to provide the comprehensive, fully functional, and fully interoperable solution that users expect in smart homes of the future.

5.2   Operational modes

This section discusses and categorizes some of the known operational modes for content management in home networks. These operational modes have been evolving during the last 10 years, adapting to the changing environment of modern mobile devices.

The home network is understood as a collection of recognizable devices capable of connecting over technologies like Wi-Fi, Ethernet, Bluetooth, etc. Consequently, from the user's perspective, the process of finding content and playing content in a home network is directly related to how to use media devices. We distinguish three types of nodes or endpoints in a media-centric home network (see Figure 5.1):

- A **source node** represents the unit that appears as the source of content in a home network. A source unit can be located within the home network or outside the home network, for example, as an Internet server.
- A **selector node** represents a helper unit with which users can navigate content catalogs. The selector, for example, can be implemented in a phone as an app that navigates the content catalog known as YouTube.
- A **target node** represents the unit that renders content in a home network.

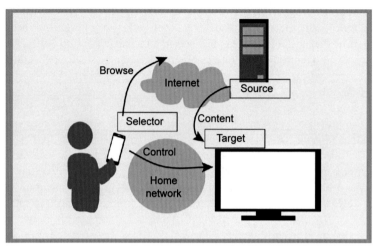

Figure 5.1 Endpoint types for content selection and playback in home networks.

Figure 5.1 illustrates the three types of endpoints that typically interact in the context of media-centric home networks. The figure shows a content–source device located in the Internet, but a source could be placed within the home network. Similarly, it is entirely possible for the selector device to be located in the Internet (e.g., Internet methods to control home devices), but the most common case is for the selector to be located at home. The target device is shown as being a TV connected to the home network, but in more complex scenarios, the target device could also be a device in a network outside the local network (e.g., TVs that can be remotely controlled from other locations).

Furthermore, a single device like a phone or tablet can implement simultaneously one or more of these roles – for example, a phone can store content in memory and implement multiple roles via apps. One application on the phone can expose content to the network (source node), while another application can be used to find and select content (selector node). Finally, a third application can implement the target node that plays content from the same or from other devices.

From the perspective of content management, there are multiple scenarios applicable in the context of home networks. Some examples include:

- A user finds content in one device, sourced from the same or from other devices, and plays the content on the same or in other devices (content consumption scenarios)
- A user takes content from one device and transfers the content to other devices for temporary or permanent storage (content download or upload scenarios)
- A user plays content on one device while simultaneously interacting socially with friends in a social group on the same device (convergence of media and social experiences)
- A user watches a movie or TV program on one device and simultaneously receives complementary information on a second device (second-screen scenarios)
- A user prints content from any device into any printer in a home network

From these categories, this article concentrates on the first one because, historically, it is the first important scenario for deploying content management at home.

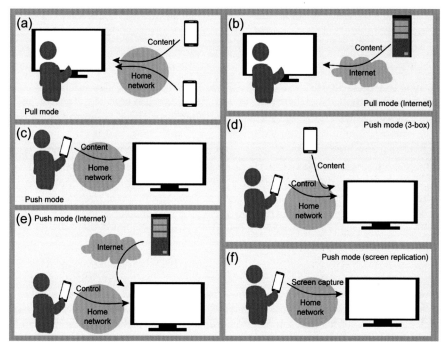

Figure 5.2 A set of common operational modes for content experiences in a home network. (a) Pull mode, (b) Pull mode (Internet), (c) Push mode, (d) Push mode (3-box), (e) Push mode (Internet), (f) Push mode (screen replication).

For the first category of content consumption scenarios, there are at least six baseline modes for selecting and playing content (see Figure 5.2):

1. **Pull mode**: User interacts with the user interface of a target node with the purpose of bringing and playing content from a source device located somewhere else in the home network. Content is transferred from the source to the target node for immediate consumption.
2. **Pull mode (Internet)**: User interacts with the user interface of a target device with the purpose of playing content from a source located outside of the home network, typically a server in the Internet. Content is transferred from an Internet server to the target device for immediate consumption.
3. **Push mode**: User interacts with the user interface of a selector with the purpose of choosing content from the same device and playing the content on a target located somewhere else in the home network.
4. **Push mode (3-box)**: User interacts with the user interface of a selector with the purpose of choosing content from a source located somewhere else in the home network, and playing the content on a target also located somewhere else in the home network.
5. **Push mode (Internet)**: User interacts with the user interface of a selector device with the purpose of choosing content from an Internet server and playing the content on a target located somewhere else in the home network. In this case, there are two additional configurations:
   a. Direct mode: The selector sends the target device a URL for the content. The target device retrieves and plays the content directly from the Internet source.
   b. Proxy mode: The selector retrieves content from the Internet source and sends the content bytes to the target device.

6. Push mode (screen replication): User interacts with the user interface of a selector device with the purpose of replicating the device's screen on a target device located somewhere else in the home network.

The different modes described here use terminology that reflects the user's perception. In pull mode, the user perceives the experience of transferring content as pulling content from somewhere else. In push mode, the user perceives the experience of transferring content as pushing content to somewhere else. The terms "push" and "pull" described here are unrelated to similar nomenclature used for the underlying content transport protocol. In fact, it is entirely possible to implement the push mode using a pull transport protocol like HTTP.

Scenario (b) described as pull mode (Internet) corresponds to the simple case of a client-server architecture. The user browses content from an Internet source using a browser or an app in a client device. The content is then transferred from an Internet server to the device for playback. This scenario is undoubtedly the most popular in terms of usage of all scenarios described here.

Other than scenario (b), all other scenarios require device communication protocols over a home network. Scenario (a) represents the first topology that the industry attempted to popularize for using media over a home network. In scenario (a), the user interacts with a target node to bring content from connected source nodes. Microsoft implemented one of the first source nodes using the Universal Plug and Play (UPnP) specifications in Windows Vista. Sony introduced one of the most popular target nodes also using the UPnP/DLNA specifications as part of PlayStation 3. A majority of TVs implementing the DLNA standard can be used as target nodes for scenario (b).

Although there is an important community of users very comfortable with devices operating in mode (b), the average home network user rarely sets up devices to work under this configuration. One of the reasons for this model failing to get significant traction among average users seems to be that it no longer matches the content accumulation patterns of modern-day users.

Ten years ago, the industry assumed that users would create and manage digital libraries with pictures, videos, and music, and that these libraries would be stored in computers or in specialized network-attached storage (NAS) devices. However, this usage pattern no longer effectively reflects how users organize and manage content today.

Today, users accumulate large numbers of content items, but they do not usually create well-organized digital libraries in computers or NAS devices. Instead, users keep their personal content distributed among their personal mobile devices (phone, tablets). Users also share their content using social distribution services so that content is also available to friends and family for consumption in other mobile devices.

Because of the prevalence of accessing content via mobile devices, the relevant topology for personal content experiences in home devices has shifted from the pull mode to the push mode. Users interact with their mobile devices (selectors) to choose some content. The content can be stored locally or can be available in distribution Websites (social networks) or in other devices in the home network. Once users select content with their mobile devices, they can redirect playback to some target device in the network.

The six modes described here represent some of the baseline topologies for content management. It is possible to design more complex scenarios, especially if we

Table 5.1 **Operational modes in currently available technologies for content management in home networks**

Technology	Operational modes
DLNA	Pull mode, Push mode, Push mode (3-box),
Samsung AllShare	Pull mode, Push mode, Push mode (3-box), Push mode (screen replication)
Apple AirPlay	Push mode, Push mode (3-box), Push mode (Internet), Push mode (screen replication)
Google Chromecast	Push mode (Internet)

consider availability of multiple sources, selectors, and targets. For example, if a user uses the push mode to send music to N target devices, then the following additional operational modes become relevant:

• Content is transferred to each of the N target devices without playback synchronization
• Content is transferred to each of the N target devices with the goal of synchronized playback
• If the N devices are grouped in zones (N_1 devices correspond to zone 1, N_2 devices correspond to zone 2, and so forth), then different content items can be pushed per zone. All devices in a zone play the content using synchronized playback. This is the so-called party mode for networked content.

Table 5.1 shows some of the technologies available today for content management and the type of scenarios/modes currently available in each of the technologies.

5.3 Technology analysis

In terms of technical descriptions of current content consumption over home networks, there are common functions to most of the currently available solutions. In all the solutions, we can distinguish – with different degrees of sophistication – the following components:

• Device Discovery
• Service Discovery
• Content Selection
• Content Transport
• Content Control.

5.3.1 Device and service discovery

If a device connects to a home network, the device needs to discover if there are other devices also connected to the network. The procedures that enable this function are known collectively as device discovery.

If there are other devices in the home network, the connecting device needs to determine the type of media services (if any) offered by such devices. The procedures that enable this function are known collectively as service discovery.

For device and service discovery, Apple's AirPlay uses the multicast DNS protocol known as mDNS (RFC 6762, 2013). mDNS is one of the protocols in a collection of zero-configuration protocols used by Apple devices and known collectively as Bonjour (Apple Inc., 2013).

DLNA implementations use the UPnP Device Discovery and Description protocol (UPnP DA, 2008). Chromecast applications use a protocol published by Netflix and known as Discovery And Launch protocol (DIAL). For device and service discovery, the DIAL protocol also uses UPnP (Netflix Inc., 2014).

The two protocols, mDNS and UPnP, provide very similar functionality. Although the protocols are different in their structural components, the underlying principles are similar. In both cases, devices and services advertise their presence, sending information on a multicast address. In both cases, any device that joins a network can send a multicast query to ask if certain types of services are available. In both cases, if a device includes the requested service, the device responds to the query with a response that includes service information.

An advantage of the mDNS protocol is that it includes name resolution. Devices and services can select a unique friendly name that serves as a network identifier. In the UPnP protocol, devices and services use a long random number known as UUID as an identifier. An advantage of the UPnP protocol is that device and service description use a rich XML format, allowing expressive descriptions for configuration parameters and other relevant information.

Figure 5.3 illustrates the main procedures used by a device to discover other devices and services using either mDNS or UPnP. The top diagram shows a device that joins a home network sending information about itself and its services. The bottom diagram

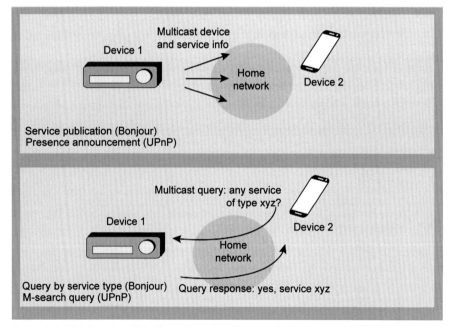

Figure 5.3 Multicast-based device and service discovery in Bonjour and UPnP.

shows a second device making a multicast query to find a particular service in the network. In this case, device 1 responds affirmatively to the request. At this time, device 2 can invoke specific functionality on the service found on device 1.

5.3.2   Content selection

In the push or pull models, the user interacts with one device and selects content from a pool of available content. This process is known as content selection. Content selection can be as simple as navigating the files stored in a device or navigating videos on a Website, but in some cases it presents additional complexities. For example, not all files stored in the file system can be carried over a network for playback in target devices. System designers need to decide if they hide these files from the content selection process, or if they show these files to the user but add a warning if the user tries to stream the content over the network.

In the pull mode (mode "a") and in the 3-box push mode (mode "d"), the user selects content exposed by a second device in the home network. A device that exposes content to other devices in the network could use web technologies for this purpose. The device can implement an HTTP server that serves HTML content to any requesting device. Using HTML5, for example, a device could expose a page with lists of video items and the required content URLs for streaming and playing.

The web server method described in the previous paragraph results in a solution compatible with Internet models (such as modes "b" and "e"). However, this solution requires the source device to use an identifiable domain name within the local network. Home networks do not normally have a DNS server managing local domains; hence, a separate naming protocol like mDNS becomes necessary.

Instead of implementing a naming service, in the DLNA approach, devices implement a service known as the content directory service (CDS). This service is used to expose content lists to other devices in a home network. Any connected device can request a CDS service in the source device to return a list of available content according to some filtering and/or search criteria. The CDS returns the requested content list using an XML format. If a user uses a phone as the selector device, the phone can use CDS request/response messages to obtain content information for content available in source devices.

Independently of using HTML5 or a CDS, the requesting device typically needs to know three important pieces of information for each content item: the type of content, the location identifier for retrieving the content, and the transport protocol that should be used to retrieve the content.

In the HTML5 case, a browser can determine the type of content examining two data fields in an HTML5 page: the file extension and the MIME type. Although not very common, an HTML5 MIME type declaration can contain additional codec information as parameters in the MIME type. An example of a video item declaration including codec information is

```
<video>
    <source src="myMovie.mp4" type="video/mp4; codecs='avc1.4D401E,
    mp4a.40.2'">
</video>
```

The MIME type indicates the multiplexing or file format for a content item. In the example, the MIME type indicates the use of the MP4 file format. The "codecs" parameter indicates the actual codecs for compressing and packaging media streams. In the example, it describes particular profiles and levels for the AVC video codec and the MP4 audio codec.

In the case of DLNA, the CDS service exposes the content type using also a MIME type. DLNA uses a separate field known as a Profile ID to provide additional description of actual encoding methods. For example:

```
<res protocol-info="http-get":*:video/mp4:DLNA.ORG_PN=AVC_MP4_BL_
CIF15_AAC_520 >
```

In this case, the <res> element describes a resource using four fields separated by a colon character. The third field is the content MIME type, which in the example is video/MP4. The fourth field carries, among other things, a "DLNA.ORG_PN" parameter. This parameter exposes the Profile ID, which is a text token identifying the video codec, audio codec, file format, and any other major constraints such as maximum resolution and bit rates. In the example, the Profile ID indicates the use of AVC video, with AAC audio, in an MP4 file. The maximum resolution is CIF15, and the maximum bit rate is 520 kilobits per second (kbps).

Consequently, regardless of the protocols used, as a result of the content selection process, the selector device knows:

- The URL of the selected content
- The MIME type and codec information for the selected content
- The transport protocol to move content bytes across the network. This protocol is typically extracted from the content URL schema, such as HTTP.

5.3.3   Content transport

The previous section indicates that a device in the network exposes content using methods such as HTML5 or a CDS. In HTML5, the "src" attribute carries a URL pointing to the content location. The URL includes implicitly or explicitly a schema declaration that identifies the transport protocol. For example:

```
<video>
   <source src="myMovie.mp4" type="video/mp4; codecs='avc1.4D401E,
   mp4a.40.2'">
   <source src="http://somedomain.com/store/xyzmov.mp4"type="video/
   mp4">
</video>
```

In the above declaration, the same video is available from two locations. The second location is identified using an absolute URL including explicitly the schema identifier ("http"), which specifies HTTP as the transfer protocol for the video content (RFC 2616, 1999). The first location uses a relative URL. In this case, the browser uses the schema and base path of the HTML document to construct an absolute URL for the video source. Because "http" is the default transfer protocol for documents in HTML, then in this second case "http" is also the transport protocol for the video content.

In the UPnP/DLNA protocols, any content declaration uses XML and follows this pattern:

```
<item>
<res protocol-info="http-get":*:video/mp4:DLNA.ORG_PN=AVC_MP4_BL_
CIF15_AAC_520 >
http://192.168.1.25/store/xyzmov.mp4</res>
<res protocol-info="http-get":*:video/mpeg:DLNA.ORG_PN=MPEG_TS_SD_
EU_ISO >
http://192.168.1.27/store/xcode?mode=21</res>
</item>
```

The video item described in this XML fragment can be accessed from two locations in two different formats. The first field in the "protocol-info" attribute specifies the transport protocol, which in this case corresponds to the HTTP GET method.

In protocols like HTML5 and CDS, it is possible to declare transport protocols other than HTTP; for example, implementations can use RTSP/RTP. However, in practice this option is rarely used.

If a document is transferred from source to target using HTTP, the entire document needs to be received before the target device can use the information. However, in the case of audio and video streams, it is possible for a target device to start decoding and playing content using only an initial fragment of the file. The target device decodes and plays content while it receives new content packets from the network. This type of operation is normally called progressive download.

Because audio and video streams must be rendered at a specific rate (e.g., a rate of approximately 30 frames per second for some video streams) the connection channel between the source and the target must be capable of sustaining a particular transfer rate. If at any time the channel cannot sustain the necessary transfer rate, then the target device receives an insufficient amount of data for playback (data starvation). This causes the typical temporary video freezes during playback.

In progressive download the target device first accumulates a few seconds of content in a buffer. The target device then extracts content from the buffer for rendering. The content buffer at the receiving end prevents temporary data starvation. A large buffer tolerates longer periods of data starvation, but it results in longer delays at the time of starting playback.

Although the buffer is often designed to store just a few seconds of video, this delay can be unacceptable for certain scenarios like real-time communications (RTC). A more modern scenario has emerged recently in content management where users physically interact with devices to transfer and play content. For example, a user may tap a TV with a phone to transfer a video file for immediate display (see Figure 5.4).

These cyber–physical interaction scenarios require zero delays to initiate playback. When we interact physically with a device or with any object in the environment, our senses naturally expect immediate reactions. Furthermore, unlike some of the RTC solutions, video transfers between devices should not reduce or compromise video quality. Users

Figure 5.4 In cyber–physical interactions (e.g., a user taps a TV with a phone to transfer video), the playback process must start with virtually zero delay without compromising the high quality of the transferred video.

expect the high-quality content available in their mobile devices to play with the highest possible quality in their large-screen TVs. Consequently, cyber–physical interactivity requires zero startup delays and the transfer of high-quality bandwidth-intensive content.

It is difficult to satisfy both requirements using conventional content transfer architectures such as those used in DLNA or in any of the competing alternatives. In the conventional systems, an HTTP transfer is initiated as soon as the user decides to play content in a target device.

The contextually proactive systems discussed in Section 5.4 have the ability to reduce startup delays to almost zero without compromising content quality. Hence, contextual proactivity constitutes a reasonable option to enable future content management scenarios that rely on more intuitive cyber–physical interactions.

5.3.4 Content control

The typical control operations for media devices include play, pause, stop, seek, fast forward, and rewind. A content management architecture that supports some of the scenarios described in Figure 5.2 must enable some or all of these control actions.

In the pull scenarios (modes "a" and "b" in Figure 5.2), the user interacts with the user interface in the target device to control playback in the same target device. Hence, the implementation of control functions is relatively easy. The user interface in this case sends control actions directly to the underlying software and/or hardware that manages playback. The underlying software interacts with the source device by temporarily pausing or stopping the transfer of content bytes (for pause or stop actions), or by requesting nonconsecutive content fragments (for seek, fast forward, or rewind).

In the push scenarios (modes "c" through "f" in Figure 5.2), the user interacts with the user interface in the selector device to control playback in the target device. Hence, the implementation of control functions requires a protocol to transfer control requests from the selector to the target device. Upon receiving control requests, the target device interacts with the source device to control the streaming of content. As before, the streaming between a source and a target can be paused or stopped (for pause and stop actions). The target can request nonconsecutive content fragments (for seek, fast forward, or rewind). For example, the DLNA architecture implements push modes "c" and "d". A selector node is known as a digital media controller (DMC) while a target node is known as a digital media renderer (DMR). DLNA defines an open protocol that allows DMC devices to send control commands to DMR devices. In the protocol, a DMC device sends HTTP POST messages to the target DMR devices. The payload in these POST messages carries information about the actual control command (play, stop, pause, or seek) that should be executed.

The implementation of a network protocol like HTTP POST for controlling playback in remote devices is relatively simple. There are, however, some complications at the time of implementing a user interface that enables these operations.

User interface designers that allow control commands over a network can choose from a number of options. For example:

- Option 1:
  - Step 1: User finds and selects content using the user interface in a selector device.
  - Step 2: User finds devices in the network that can play the selected content.
  - Step 3: User clicks on a "play" button (using the user interface in the selector device) to initiate playback.
- Option 2:
  - Step 1: User finds and selects target devices using the user interface in a selector device.
  - Step 2: User finds and selects content that can play in the selected device.
  - Step 3: User clicks on a "play" button (using the user interface in the selector device) to initiate playback.
- Option 3:
  - Step 1: User finds and selects content using the user interface in a selector device.
  - Step 2: User clicks on a "play" button (using the user interface in the selector device) to start playing the content in the selector device.
  - Step 3: At any time the user can change a setting for a default output device, in which case content is automatically redirected to the new output device (target).

Without some exhaustive HCI research, it is hard to say which of these options (if any) constitutes a more satisfactory experience. Device manufacturers and software vendors design both the look and feel of their user interfaces and the user interface procedures that enable network browsing and control. This is different, for example, from conventional web browsers that have different look and feel (according to the software vendor) but similar user interface procedures for the main tasks.

If a user has devices of different brands and downloads different media applications, this user now has to learn the user interface style and user interface procedures for each device and application. Technically savvy users have no problem discovering the modus operandi of a device or an application that implements networked browsing and control. However, the diversity of alternatives in current implementations can seriously encumber the experience for average users. In a world where users have a multiplicity of choices for devices and applications, any feature that requires users to go through learning each time they buy a new device or download a new application severely limits its usefulness.

5.3.5   Limitations of deployed solutions

Despite the significant progress made during the last 10 years in terms of actual implementations and mass-market deployments, the content management technologies described here have not reached the level of user acceptance, domestic ubiquity, and everyday access that practitioners originally envisioned.

There is now a growing user base that consistently uses technologies like AirPlay, DLNA, and Chromecast. These users typically set up their home networks and configure their devices so that many of the operational modes described earlier in this chapter become available. However, it is also evident that the user base is principally composed of technically savvy people curious about new developments and new features. This group constitutes what the technology industry often refers to as early adopters.

However, it is also worth noting that for any technology to be considered a success in the home environment, it must become truly domestic. Domestic technologies like electricity or televisions touch the lives of home dwellers on a daily basis. They are relatively simple and intuitive to use. Users in general can operate and interact with these technologies at any time according to their respective needs.

In the case of current content management solutions for home networks, there are several variables preventing the technologies from becoming truly domestic. These problems include constrained scopes, fragmented support for activity domains, lack of intuitive user experiences, and others.

Constrained scope – Users want to exchange content over home networks, and they want this option available for all types of content. Users want to manage content that comes from their phones (pictures, videos, music), content that comes from Internet music and video subscriptions, content from DVD players, Blu-ray players, and from their television sets, etc. In the perfect home network of the future all content types can flow from any type of source to any type of target. Any content type regardless of its origin should be managed in a similar consistent and intuitive manner. The technologies available today provide solutions for some of these content types, but fall extremely short from the comprehensive model that home users want.

Fragmented support for activity domains – Current content management technologies work consistently as long as the user operates within the particular activity domain for which the technology has been designed. This is true despite the users' strong desire to switch often from one activity domain to another. Users want to switch from the TV viewing domain to the DLNA picture-sharing domain and then switch to the AirPlay mirroring domain, and from there possibly back to TV viewing. The current technologies provide very limited support for crossing activity domains. The user needs to learn how to set up the correct HDMI port, and which input should be selected to go back to TV mode, and how to start AirPlay screen-mirroring in a phone. To reach truly domestic ubiquity, any content management solution needs to satisfactorily address the multiplicity of activity domains. In the smart home of the future, users will be capable of switching back and forth between activity domains at any time and with no effort.

Lack of intuitive user experiences – Currently, device manufacturers and software vendors provide different usage experiences for the same or a similar set of media sharing procedures. The diversity of devices and software, each with its own usability principles, creates fragmented experiences that encumber the assimilation of these technologies. This lack of intuitive homogeneous access to content management creates a steep learning curve for average users. The user has to learn the operational procedures for each device or each software package to integrate the device or software to a content network. This lack of intuitive and homogeneous access prevents average users from adopting and using the technologies on a daily basis – but perhaps more important, it prevents average users from diagnosing, repairing, and maintaining the network in good operational conditions at all times.

5.4 Contextual awareness and intelligence

The previous sections have summarized the significant progress made during the last decade to bring content management solutions to the home network. The previous sections also describe some of the limitations that currently block the technologies from becoming the comprehensive domestic content management solutions that users expect.

In this section, we argue that addressing the described problems requires a new approach. We argue that any type of user experiences at home, including those required for network content management, can be improved by having a network of interconnected intelligent devices, where devices have the ability to interpret user intent (UI) and decide on the proper set of actions to enable intent. These intelligent devices communicate with each other for the purpose of exchanging UI, exchanging context information (CI), exchanging rules, setting up connections, preparing the network, and preparing content. All these actions happen as a consequence of detecting UI, and they happen typically in anticipation of user needs. We use the term *contextual proactivity* to identify the design principles and methodology proposed in this work.

The notion of introducing distributed intelligence to home network devices has been investigated extensively. However, the original research targeted mainly home automation and service optimization. With these goals in mind, the research community looked at topics like location awareness, context awareness, multimodal sensors, activity detection, event automation, adaptive environments, and others (Alam, Reaz, & Ali, 2012). Despite the significant results and continuous improvements, most of this research on smart homes has remained an activity in academic and industry labs and has not crossed the line into production.

There are multiple reasons that original smart home research has not yet crossed the line into production, but two of the important reasons include scope and reliability. Research often focuses on too many large-scale scenarios rather than concentrating on a few well-defined cases (scope). Furthermore, research methodologies like machine learning, probabilistic classifiers, neural networks, and others, result in success rates that are high from a research perspective (upper 80% or 90%), but low from a domestic product perspective where success rates should be around 100%. In this sense, we agree with researchers like T. Yamazaki, who argues that future smart home research should address problems like improved interfaces that effectively link users (intent, sentiments, and situations) and the ecosystem of home devices (Yamazaki, 2006).

There is increasing recent interest in user-centric design (UCD) for human–machine interactions. In terms of smart homes, UCD is a methodology that places the user at the center of the home device ecosystem. User experience research in this field covers promising areas like mixed reality (Dooley, Davies, Ball, & Callaghan, 2010), tangible augmented reality (Billinghurst, Kato, & Poupyrev, 2008), voice interfaces (Portet, Vacher, Golanski, Roux, & Meillon, 2013), gesture-based systems (Kuhnel et al., 2011), brain–computer interfacing (Edlinger, Holzner, & Guger, 2011), wearable controllers (De Russis, Bonino, & Corno, 2013), and others. These user-centric interface systems become the ideal technologies to capture UI, which is then propagated through a network of intelligent contextually proactive agents (CPAs) for decision making and multimedia management. The next section describes the design principles for these CPAs.

5.4.1   Contextually proactive systems

We define a *contextually proactive system* as a device that has the following properties:

- The device is able to sense its surroundings collecting CI
- The device hosts applications (apps) that serve as an interface to detect or acquire UI
- The device hosts a CPA, which translates UI into rules that enable UI
- The CPA connects with peer CPA entities in the network to cooperate toward configuring the network and satisfying UI

Figure 5.5 illustrates this definition. As an example, consider the case of a user who is watching a movie on a phone while riding a train on her way home. When this user reaches home, she probably wants to finish watching the movie. At home, she has better devices to watch a movie; for example, a large-screen TV in a living room.

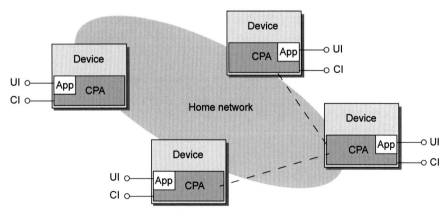

Figure 5.5 A collection of contextually proactive devices in a home network.

An application on the phone (e.g., the media player) determines that the user is watching a movie, and it can easily infer that the intent of the user is to continue watching the movie. For example, if the user pauses the video in the transition from the train station to home, the availability of a paused video is a clear indication that a user intends to continue watching the movie at a later time.

In other examples and/or cases, the application could use sensors to capture intent, and voice recognition and language processing to understand user requests; it could collect intent using modern user-centric interfaces (e.g., mixed reality, gestures, tangible augmented reality, etc.), or it could collect input from the user via conventional user interfaces.

The application sends information about UI to the CPA. The CPA agent collects CI directly from sensors or, preferably, from a separate service that may be available in the device for this purpose. Using CI, the agent determines, for example, if the user is at home or if the user is near a TV.

In the case of the user who watches a movie on a train on her way home, when the user reaches home, the CPA knows the following types of CI: the user's phone has a paused video, the user's phone is connected to the home network, the user is at home but is not near the TV, and later it knows if the user is near a TV and operating the remote control.

A rule-based inference system is used to collect all the CI and trigger actions that automatically enable sharing the movie with nearby devices at home. In this way, if the user sits in front of a TV and starts operating it, the user may get a notification in her phone (or on the TV) asking her if she would like to continue watching the paused movie. The system has proactively prepared the connections, configured the devices, and even transferred all or fragments of content in anticipation of making the question. The end-to-end system is ready for media sharing in anticipation of user needs. This is the main characteristic of contextually proactive systems. The user no longer needs to learn device or software operational procedures. Instead, the system proactively performs all the necessary tasks on the user's behalf.

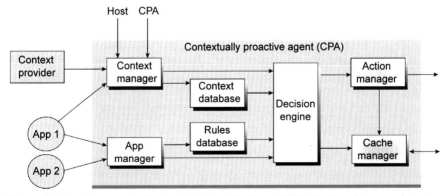

Figure 5.6 Architecture of a Contextually Proactive Agent (CPA) that translates user intent (UI) into content management actions.

Figure 5.6 illustrates the different subcomponents and main data flows in a CPA. These subcomponents perform the following functions:

Context Manager – This component collects CI from a variety of sources, including a Context Provider Service, applications installed in the device, the host device, and the same or other CPAs. The Context Manager does not acquire all information at all times. It only acquires the relevant subset required to process currently available rules.

App Manager – This component communicates with installed applications in the device. Some applications transfer UIs that need to be converted into rules. Other applications convert UIs into rules and transfer the rules directly to the CPA system. An XML-based Interchange Description Language is used for exchanging information between applications and the CPA.

Context Database – This component stores the current context data necessary to determine if some rules can be satisfied or not. The database is updated periodically.

Rules Database – This component stores all the known rules that enable contextually proactive behaviors. This database can be implemented as a deductive database capable of deriving additional rules based on inference. The Rules Database includes a set of rules that describe the environment; for example, it may include rules like "the game console is near the TV." Using this rule, the database can infer that a user who turns on the game console is also near the TV. In our implementation, this database uses Datalog for inference and rule data retrieval (Heredia, Kumar, Nishimura, Hsieh, & Messer, 2014).

Decision Engine – This component examines the CI and rules, and determines if some of the rules are satisfied by current context conditions. Whenever certain rules are satisfied by context conditions, the Decision Engine triggers actions on the system. The actions include: send actionable data to peer CPAs, receive actionable data from peer CPAs, send content (or fragments) to peer CPAs, receive content (or fragments) from peer CPAs, send notifications to one or more peer devices including the host device.

Action Manager – This component processes actionable data and orders the host and/or peer devices to execute the actions. Examples include "turn on one of the TVs," "switch HDMI," "display overlay information," "display a notification," etc.

Cache Manager – This component processes content transfer requests (sending or receiving) and content caching. Content is often transferred and temporarily cached in anticipation of user needs. The cache manager performs cache optimization over time. If multiple peer devices request caching, the cache manager handles priorities and policies. At any time the cache manager can assign cache space or remove cached objects for particular devices.

We have implemented the architecture described in Figure 5.6 and demonstrated that it can be successfully used to address multiple content management scenarios (Heredia et al., 2014). We have tested the system with five real end-to-end scenarios and more than 20 simulated scenarios. Some of the example scenarios include:

- A user pauses playback of a video on a phone. The video is either local (available on the same device) or streamed from the Internet. Whenever the user approaches the TV, the device makes an offer to the user to play the video.
- A user takes pictures with a phone. The device makes an offer to the user to display the new pictures whenever the user approaches a TV. Additionally, the offer is made only when specific people are colocated with the user near the TV.

- A user is listening to music on a phone. The music is automatically switched to a nearby output device and/or speakers when the user enters a particular room or area in a home.
- A user is playing a game on one device and reaches a certain level. The user continues playing the game from the specified level on the second device if the first device touches the second one.
- A user finds an address and directions on a map using a phone. The map application remains active in the phone. If the user enters a car and taps the dashboard with his phone, the map and directions are transferred to the car navigation system.
- A user composes a video using a phone and then requests the system to invite a second person to watch the video on a TV when this second person is near the TV at a specific date. For example, the invitation happens during the second person's birthday.
- A user is watching a certain program on a TV and periodically checks his or her phone. The TV and the user's phone co-ordinate to download proactively a second-screen app. The user can be notified of the availability of this app, and may want to run it for an enhanced experience.
- A user approaches a TV with the intention of watching some programs. The TV and the user's phone co-ordinate to download and launch proactively an app capable of controlling the TV. The user can use his or her phone to fully control the TV.

There are many other scenarios that can be implemented using contextual proactivity. In any of these scenarios, users express intent or the system detects their intent. Once intent has been determined, the collection of distributed agents proactively cooperates to set up connections and exchange content to satisfy the intent. Under the proper context conditions, the system activates content-related behaviors such as transferring rules, transferring content, playing content, triggering notifications, or transferring control.

We have implemented and tested nearly complete solutions for the first, second, and fourth scenarios in the list above. For example, in the second scenario, the discovery of new pictures in the user's smart phone causes the CPA in the phone to generate a sequence of proactive rules to be executed on the phone:

- Rule 1:
  - Conditions:
    - C11: The user's phone is in a Wi-Fi network
    - C12: The user's phone can connect with TVs in this network
    - C13: The user's phone has new pictures
  - Actions:
    - A11: The user's phone transfers new pictures to all reachable TVs
    - A22: The TV sends back a report that selected content has been cached
- Rule 2:
  - Conditions:
    - C21: The user's phone is in front of a TV
    - C22: The TV has cached all new pictures
  - Actions:
    - A21: The phone displays a notification inviting the user to view the new pictures on a TV
- Rule 3:
  - Conditions:
    - C31: The user responds "yes" to the notification
  - Actions:
    - A31: The phone sends an order to play the pictures to the TV
    - A32: A playback control UI is displayed on the phone for further control actions

This sequence of rules determines the proactive actions that the system performs on behalf of the user. In this case, the purpose of the rules is to facilitate the viewing of any new pictures on a TV. If the user has new pictures on a phone, and if the user is in front of a connected TV, the system of connected CPAs invites the user to view the new pictures on a TV. The sequence above shows a simplified version of the rules. More complex rules are necessary to cover additional options like repeating notifications, partially caching a set of images, and others.

Notice that from a user's perspective, the user simply responds yes/no to an invitation to view pictures on a TV. All users that have tested the system prototype have had a positive reaction due to the simplicity of operation. Furthermore, because the pictures are transferred in anticipation of usage, if a user accepts the offer to view the pictures, the TV shows the first picture very quickly, avoiding the usual delays caused by network transfers.

To trigger the actions described in rules, devices need to determine whether the contextual conditions are satisfied. In the case of the above example, any phone can detect easily if it is connected to a Wi-Fi network (C11). To determine if there are target devices available in the network, the implementation uses a discovery protocol (C12). To determine if there are new pictures on the user's phone, the CPA implementation checks the phone's file database (C13). To determine if a phone is in front of the TV, the current implementation uses an attenuated Bluetooth transmitter on the TV and a conventional Bluetooth sensor on the phone (C21).

The implementation of conditions that require the status of a connected device, as well as the implementation of action messages between devices, requires the use of a communication protocol between CPA agents. Our first implementation uses Qualcomm's AllJoyn as the application-discovery, message-transfer, and file-transfer protocol (AllJoyn, no date).

Contextual proactivity constitutes a different methodology to implement content management in home networks than the more classical solutions described in Section 5.3. This new methodology requires devices to implement an intelligent core that directs the exchange of rules, context, actions, and content. Table 5.2 illustrates some of the important differences between these methodologies. Table 5.3 illustrates the differences in software stack between a conventional DLNA implementation and a contextually proactive implementation.

An examination of Table 5.3 shows that while DLNA uses the notions of device and service discovery, the CPA architecture uses the notion of application discovery. Each CPA in the network behaves as an application that discovers peer applications in surrounding devices. As mentioned above, we use Qualcomm's AllJoyn stack for application discovery, message exchange, and file transfer.

Table 5.3 also shows that while DLNA relies on server-client behavior, the CPA architecture relies on peer-to-peer behaviors. In DLNA, a client device (such as a Digital Media Player or a DMC) makes requests to a Digital Media Server device. In the CPA architecture, any CPA agent can query peer agents requesting content information or playback information.

Finally, Table 5.3 shows that in conventional architectures like DLNA's, the user explicitly interacts with particular devices in the network, transferring content from one device to another. In the CPA architecture, the user interacts with the distributed

Table 5.2 **Differences in methodology between conventional solutions and contextually proactive systems**

Conventional solutions	Contextual proactivity
User needs to identify device behavior and/or purpose in terms of sources, targets, and selectors.	User interacts with any device in the network regardless of its type. All devices are considered peers.
User needs to operate menus and/or buttons to initiate and control content management operations.	User needs to express intent while devices cooperate to set up any necessary elements to enable content management.
Content transfers between devices happen the moment when a user needs to play content on certain devices.	Content is transferred proactively (in full or in fragments) to potential targets. A cache manager subsystem handles caching operations.
Users need to master content management procedures per device and/or application. If things do not work as expected, users need to consider connectivity constraints to check for problems.	Users express intent. Once all network elements have been proactively configured, users often operate the system with a "yes" or "no" response to device notifications.

intelligence formed by a cooperative of CPA agents. The CPA agents interpret UI, configure some operational rules, distribute decision-making rules throughout the network, and track context conditions to evaluate the status of stored rules. Using this architecture, CPA agents can effectively use a notification system to inform the user about entertainment and content management options at home.

5.5 Conclusions

This work describes the status of content management technologies in home networks. It describes the recent industry efforts to deploy successful solutions, the characteristics and principles of deployed solutions, and their key functionalities and device architectures. After more than a decade of work in this area, these deployed technologies offer reasonable solutions to particular cases within well-defined operational scopes, but no comprehensive solution capable of managing all types of content and all types of devices exists today. Furthermore, the deployed technologies remain an attractive proposition for technically inclined users but have not yet acquired the simplicity, flexibility, and cross-domain functionality required to capture the hearts and minds of the average user.

This work highlights the limitations of existing solutions and argues that the deployment of more intuitive systems requires investments in device intelligence. This work introduces the notion of *contextual proactivity* as a methodology to build device intelligence. Unlike existing industry solutions that require users to master procedures for content management, the proposed alternative relies on devices capable of interpreting UI, sensing context, and proactively setting up all network elements toward satisfying UI.

Table 5.3 Differences in software protocols and implementations between Digital Living Network Alliance (DLNA) devices and a first implementation of a contextually proactive device

DLNA device architecture	Contextually proactive agent (CPA) device architecture (initial prototype)
Uses simple service discovery protocol (SSDP) for network discovery. Devices discover other devices and services.	Uses peer-to-peer protocols for device and application discovery. CPA agents are considered applications. Applications discover other applications.
Uses XML-based descriptions to expose and declare services (service description document).	The discovery protocol allows any app to discover a peer app. Any app can query a second app requesting information about services and capabilities.
Content selection uses a content directory service (CDS) for discovering servers and their content.	Any device can query peer devices to determine content availability using conventional file/folder read procedures.
Requires classification of devices as servers, controllers, players, and renderers. Requires specific protocols for interoperability of these device classes.	All devices behave as an integrated network of semi-intelligent agents (CPAs). These agents operate in a peer-to-peer configuration. Although devices can have different capabilities, all agents have similar roles.
The main types of interactions are between a user and specific classes of devices (mainly servers, renderers, and controllers).	The main types of interactions are between a user and the network of semi-intelligent agents. These agents cooperate to address user requests.
Uses HTTP progressive download (streaming) as the method to transfer content from source to destination. The transfer is often triggered upon user request.	Uses conventional file transfer over a network. Because the system uses proactive transfers, the exchange often happens in anticipation of user needs triggered by proximity context conditions.
Actions always triggered by user interactions with controllers.	Actions triggered by intelligent agents that make decisions on behalf of the user (based on sensed context conditions).

We describe the architecture of contextually proactive devices in a home network. A device of this kind includes functionality to process UI and to distribute actions and operational rules to other devices. A device of this kind also includes a deductive database of rules that can be used to infer contextual knowledge. At any time, devices acquire CI and determine the subset of rules that should be triggered. These distributed rules are used to automatically set up connections, proactively transfer content, and notify users about availability of content management options. Contextual proactivity offers the possibility of designing future systems with much more intuitive interfaces and simplified user operations.

References

Alam, M. R., Reaz, M. B. I., & Ali, M. A. M. (2012). A Review of smart homes—past, present, and future. *IEEE Transactions on System, Man, and Cybernetics—Part C Applications and Reviews*, 42(6), 1190–1203.

Alljoyn (no date). A common language for the internet of everything (Online). Available from: https://www.alljoyn.org/ Accessed 05.09.2014.

Apple Inc. (2013). *Bonjour overview* (Online). Available from: https://developer.apple.com/library/mac/documentation/Cocoa/Conceptual/NetServices/NetServices.pdf.

Billinghurst, M., Kato, H., & Poupyrev, I. (2008). *Tangible augmented reality*. ACM SIGGRAPH ASIA.

De Russis, L., Bonino, D., & Corno, F. (2013). The smart home controller on your wrist. In *Proceedings of the 2013 ACM conference on pervasive and ubiquitous computing adjunct publication*. ACM.

DLNA. (2003). *Exhibit C statement of founding principles* (Online). Available from: http://www.dlna.org/docs/dlna-contributor-membership-documents/dlna_organization_founding_principles.pdf?sfvrsn=2.

DLNA. (2013). *DLNA market overview report* (Online). Available from: http://www.dlna.org/dlna-for-industry/newsroom/parks-associates-report-2013/?utm_source=DLNA&utm_medium=press_release&utm_campaign=10yearanniversary.

Dooley, J., Davies, M., Ball, M., & Callaghan, V. (2010). Decloaking Big Brother. In *Proceedings of the 6th International conference on intelligent environments, Kuala Lumpur, Malaysia*.

Edlinger, G., Holzner, C., & Guger, C. (2011). A hybrid brain-computer interface for smart home control. In *Human-computer interaction, interaction techniques and environments, lecture notes in Computer Science*. Springer.

Heredia, E. A. (2011). *An introduction to the DLNA architecture: network technologies for media devices*. John Wiley & Sons.

Heredia, E. A., Kumar, S., Nishimura, J., Hsieh, G., & Messer, A. (2014). Contextual proactivity for media sharing scenarios in proximity networks. In *Proceedings of the IEEE consumer communications & networking conference, Las Vegas, Nevada, USA*.

Kühnel, C., Westermann, T., Hemmert, F., Kratz, S., Müller, A., & Möller, S. (2011). I'm home: defining and evaluating a gesture set for smart-home control. *International Journal of Human-Computer Studies*, 69(11), 693–704.

Netflix Inc. (2014). *Discovery and launch protocol specification, version 1.7* (Online). Available from: http://www.dial-multiscreen.org/dial-protocol-specification.

Portet, F., Vacher, M., Golanski, C., Roux, C., & Meillon, B. (2013). Design and evaluation of a smart home voice interface for the elderly: acceptability and objection aspects. *Journal of Personal and Ubiquitous Computing*, 17(1), 127–144.

RFC 2616. (1999). *Hypertext transfer protocol—HTTP/1.1*. Internet Engineering Task Force (IETF).

RFC 6762. (2013). *Multicast DNS*. Internet Engineering Task Force (IETF).

UPnP DA. (2008). *UPnP device architecture 1.0*. UPnP Forum.

Yamazaki, T. (2006). Beyond the smart home. In *Proceedings of the International conference on hybrid information technology, Jeju Island, Korea*.

Case study of an ecological, smart home network: iZEUS-intelligent Zero Emission Urban System

M. Isshiki[1], M. Umejima[2], M. Hirahara[3], T. Minemura[4], T. Murakami[5], S. Owada[6]
[1]Kanagawa Institute of Technology, Atsugi, Japan; [2]Keio University, Minato, Tokyo, Japan; [3]Steering Committee of ECHONET Consortium and Toshiba Corporation, Minato-ku, Japan; [4]Toshiba Corporation, Minato, Japan; [5]Technical Committee of ECHONET Consortium and Panasonic Corporation, Osaka-fu, Japan; [6]Sony Computer Science Laboratories, Inc., Shinagawa-ku, Japan

In this chapter, the example of realization under market expansion by the open platform using ECHONET Lite is introduced as a development example of a smart home network. This chapter also introduces the latest trends in ECHONET Lite and discusses the importance of converting the business into an open platform model, which is crucial to make it applicable in the international market.

We describe commercialized systems for the smart home market in Asia, especially explaining the deployment power of the open standard (ECHONET Lite)-based business. ECHONET Lite is an open network protocol, which can be beneficial to society in the twenty-first century.

Since the types of network-connected devices are increasing and home network services are diversifying, there is a growing need to make it easier to build home networks. It was therefore decided to develop ECHONET Lite, which is a transport-free network protocol so that developers can themselves select communication media that they want to use. In 2011, ECHONET Lite brought innovation to EMS. In terms of a power meter, since JSCA (government-industry collaboration in Japan) decided to adopt ECHONET Lite over IPv6 as the standardized communication interface to a power meter, all of Japan's residential power meters, exceeding over 48 million, have been functionalized as a home appliance, delivering electric consumption data to a home: voltage, ampere, watt-hour, log-data.

ECHONET Lite is designed to be easier for home network system builders and service system developers. Not only network protocols but also the definition of control commands are very important for the interoperability between multivendors. Therefore, ECHONET Lite has defined control commands of about 90 devices to realize multivendor interoperability. Typical devices are air conditioners, lighting, fuel cells, photovoltaics, storage batteries, heat pump water heaters, EV/PHV, and smart meters.

The following five themes are explained.

1. Overview of smart home businesses today
2. Overview of ECHONET Lite (International Open Standard for Home appliances and utilities)

Ecological Design of Smart Home Networks. http://dx.doi.org/10.1016/B978-1-78242-119-1.00006-0

3. Overview of smart house businesses 1: Panasonic Activities
4. Overview of smart house businesses 2: Toshiba Activities
5. Open source software and related activities for ECHONET Lite with web technologies.

6.1 Overview of smart home businesses today (by Isshiki and Umejima)

6.1.1 Smart home and ECHONET Lite

As the name implies, the term *smart home* refers to "smart" or "intelligent" houses. Its objective is to provide a home energy management system (HEMS) that manages energy consumption within homes. HEMS is an energy consumption management system that maximizes electricity efficiency within homes and optimally controls home equipment that is used to reduce and maximize energy use, as well as create and store energy.

Additionally, we feel it is important that we contribute to society by standardizing electricity and creating smart homes (HEMS) that can be linked to community energy management systems and building energy systems (BEMS).

ECHONET Lite was established by the ECHONET Consortium (ECHONET Consortium, http://www.echonet.gr.jp), a Japanese corporation with a 15-year history, and is one of the International Standard Communication protocols within HEMS. Communication protocols are one of the mutually agreed-upon rules when using computers through networks. Making ECHONET standards open to the general public will help accelerate the conversion of the smart home business into an *open platform* business model. An open platform will become an important key word for future smart house market expansion, and will be covered in this discussion.

6.1.2 ECHONET Lite latest works

Japan's Ministry of Economy, Trade and Industry formally decided to make ECHONET Lite "the recommended public standard interface for home equipment and between HEMS and smart meters." This epoch making start, introduced at a large ministry-related industry conference, was kicked off by industry-related companies who were strongly determined to "reach a decision and move forward" to alleviate social problems created by tightened energy demand following the East Japan Earthquake disaster in March 2011.

We will cover the background behind this decision and the applications for ECHONET Lite in this chapter.

1. Decision by the Energy and Smart House Standardization Investigative Council to implement HEMS

The Smart House Standardization Investigative Council was established on February 24, 2012, as a result of the November 2011 Energy/Environment Conference decision to implement an investigation to standardize smart meters and HEMS. The investigative

results can be viewed on the following website: http://www.meti.go.jp/press/2011/02/
20120224007/20120224007.hml.
JSCA committee reached the following two main decisions:

1. To implement HEMS and recommend ECHONET Lite as a standard interface for home
   equipment and between HEMS and smart meters (see Figure 6.1).
2. To promote ECHONET not only for the domestic market but also international markets and to
   establish international standardization to facilitate overseas business development.

In accordance with the above two decisions, the JSCA committee decided to
"popularize the use of smart meters by electric companies" and "adopt ECHONET
Lite as a standard within the HEMS promotion initiative." There are three main
purposes for recommending ECHONET Lite as an official standard interface:

The first is to make energy conservation possible through smart meter data acqui-
sition, visualization, and use for controlling devices. The second is to create and provide various services through making interconnectivity possible between equipment
of different manufacturers. The third purpose is to remove market entry barriers for
small and medium venture companies newly coming into the market that previously
had to adapt to different communication standards for each manufacturer.

It is also necessary to push forward with development of a global standard for the
many manufacturers that are expanding their businesses worldwide. Currently, HEMS
and smart meters are becoming the core of the smart home, and global competition
regarding their standards is increasing. Although there are different approaches, some
examples are the SEP 2.0 (Smart Energy Profile 2.0) in the United States and KNX
(KONNEX) in Europe. Regarding energy equipment with detailed control functions,
at this time Japan ECHONET Lite is one of the most highly technically advanced.
It has three strengths, which are "adapting to Internet protocols (IP)," "many numbers
of commands adapted for over hundreds of devices" and already "open and fully standardized at IEC." For this reason, it is important to continue taking advantage of these
strengths and integrate and link ECHONET Lite with world standards and specifications.

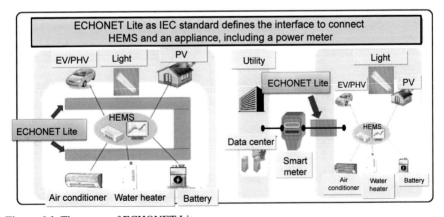

Figure 6.1 The scope of ECHONET Lite.
Interim meeting report from Smart House Standardization Investigative Council.

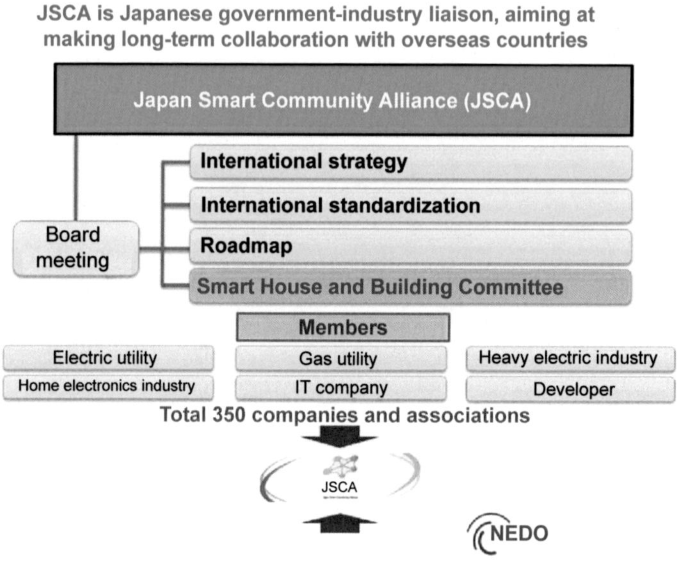

JSCA is Japanese government-industry liaison, aiming at
making long-term collaboration with overseas countries

Figure 6.2 shows the organization chart described below.

JSCA is Japanese government-industry liaison, aiming at making long-term collaboration with overseas countries

Japan Smart Community Alliance (JSCA)

Board meeting

- International strategy
- International standardization
- Roadmap
- Smart House and Building Committee

Members

| Electric utility | Gas utility | Heavy electric industry |
| Home electronics industry | IT company | Developer |

Total 350 companies and associations

JSCA → NEDO

Figure 6.2 Organization chart of JSCA.
Materials from the first meeting of JSCA.

ECHONET Lite is a communication protocol established by the ECHONET Consortium. Its standards were established on August 11, 2011, and have been publicly disclosed on the Website since December 21, 2011. Making these standards open and public has especially helped drive the business forward. This groundbreaking initiative has not only drawn attention from overseas, but it has also been effective in facilitating business collaborations; the details follow in the next section.

2. Establishment of the JSCA Smart House/Building Standards/Business Promotion Investigative Council

Against the agenda of the report by the Smart House Standardization Investigative Council (Figure 6.2), the Smart House/Building Standards/Business Promotion Investigative Council was established within the Japan Smart Community Alliance (JSCA) as the venue to implement a progress schedule and to investigate each agenda. The following was decided at the first Investigative Council meeting held on June 22, 2012:

We will first outline the eight most important types of HEMS (ECHONET Lite) energy-related equipment, which have high priority and connectivity with HEMS (http://www.meti.go.jp/committee/summary/0004668/pdf/014_03_00.pdf). These are commonly used as main home electronics and equipment in using energy efficiently and fall into the following categories (Figure 6.3):

1. Smart meters
2. Solar power equipment
3. Batteries
4. Fuel cells
5. Battery or electric-powered or hybrid cars

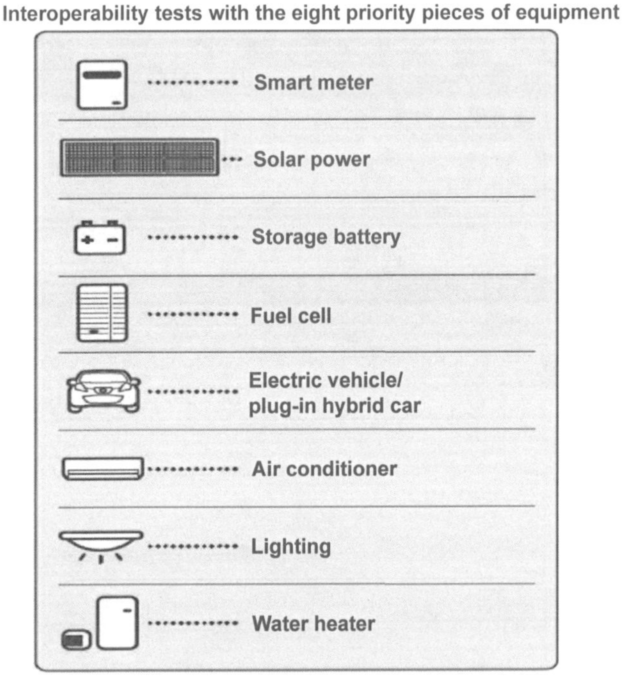

Interoperability tests with the eight priority pieces of equipment

Figure 6.3 List of eight most important types of equipment.

6. Air conditioners
7. Lighting appliances
8. Water heaters

Because all eight types of equipment produce, store, and consume high amounts of energy, it is important that they maximize energy efficiency. Regarding network configuration and IP compliance, it is recommended to use one of these three types of communication equipment – 920 MHz type, Wireless LAN, and PLC – which are especially for smart meters. In addition to outlining the eight types of equipment that have high interconnectivity with HEMS, there was a need for an environmental facility where even equipment from different manufacturers could be linked and tested for interconnectivity compatibility, which resulted in the decision to establish an examination center where ECHONET Lite authentication tests could be carried out. This examination center was named the "HEMS (ECHONET Lite) Certification Support Center" and opened on November 21, 2011. It is located within the campus of Kanagawa Institute of Technology University (see Figure 6.4).

3. Establishment of Working Guidelines between HEMS and Smart Meters in homes

The guidelines for HEMS and smart meters were established at the third JSCA Smart House/Building Standards/Business Promotion Investigative Council held on May 15, 2013, and were made public at http://www.meti.go.jp/press/2013/05/201305 15004/20130515004.html.

Figure 6.4 Exterior of home energy management system (HEMS) (ECHONET Lite) Certification Support Center.

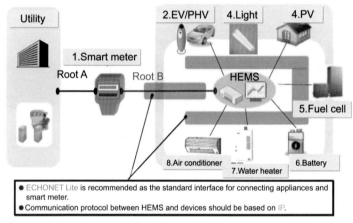

Figure 6.5 Focusing interface area of HEMS.
From HEMS-Smart Meters (B route) Working Guidelines (Publication 1.0).

It was decided at the Investigative Council to provide an experimental environment to test HEMS smart meters for home use by the HEMS Certification Support Center. The center will have available smart meters from each company so that companies will be able to bring in HEMS controllers and carry out interconnectivity compatibility experimentation (see Figure 6.5).

4. HEMS Certification Support Center

The HEMS Certification Support Center offers three main activities:

1. The first is an experimental environment for companies involved in ECHONET Lite equipment development to be able to carry out third-party authentication experiments at any time. While large companies can establish experimental environments for their equipment, small or medium venture companies cannot easily make investments. For this reason, we can

provide an open environment for companies to carry out certification testing. Providing both an open platform foundation and an open testing environment will help open up new business opportunities for newcomers. We also provide an ECHONET Lite standard software (the "most accurate golden code") in the Authentication Test Center that connects one-on-one with developed equipment, and the system detects any problems or bugs.

2. The second is the distribution of a free ECHONET Lite equipment software development kit (SDK). This has very broad applications, as there are many needs among those who want to develop technology and provide new services and products. It is important to create growing interest in the open platform business and help people get started in the business. We also have available popular computer codes like C and Java in order to make it easy for many to experiment with software development.

3. The third are activities to support the establishment of development guidelines to meet the needs of small and medium businesses and venture businesses expected to enter into the smart house/HEMS business. We are also developing ECHONET Lite educational programs and manuals. This information is made public at http://sh-center.org/.

The HEMS Certification Support Center is a facility that provides an environment where people can get hands-on practice in the open platform business. The next section will cover the concepts and objectives of the open platform business.

6.1.3 An open platform business is key

The word *open* means an environment created by a "public and using standard interface." Establishing an open platform business environment means making the environment into a platform that anyone can use and also where various business opportunities can emerge. We have seen an increase in many local specific services on a regional level. Small and medium businesses in local areas are able to create new services and businesses by taking advantage of their strengths.

The exclusive cell phone environment is one good example of how a highly manufacturer-specific service environment (sometimes called "Galapagos cell phones") has changed to a standard platform (smart phones) service environment resulting in an increase of users by service coverage. Since the smart house technology is also a specific service environment, creating manufacturer-specific service strategies will only prevent companies from competing on a global level. It is important to establish a platform environment that includes standardization, and in addition, to convert it into an "open platform business."

6.1.4 The future of the smart house

As discussed, the key to growth in the smart house market is converting it into an open platform business. ECHONET Lite is the foundation of the open platform and holds a very important position in the future of the smart house market. The Japanese government is also focusing on cultivating the smart house business. For example, on May 24, 2013, the headquarters for the Promotion of Advanced Information and Communications Society located in the Cabinet IT Strategy Headquarters declared that it would "create Japan as one of the world's most advanced IT countries." The declaration stated, "The role of aggregators is important to provide effective energy management service" and that "Efficient energy management will be achieved by establishing aggregators as a new business area" (see http://www.kantei.go.jp/jp/singi/it2/kettei/pdf/20130614/siryou1.pdf).

In the future, smart meters and HEMS will play a role in providing electricity usage data, which will result in the emergence of various new services. We especially feel that the cultivation and development of HEMS integrators who provide services to home dwellers will be a fast-growing business in local regions based on the need to provide region-specific services in these areas. This is a new business where we can expect a residential venture business to emerge from the regional housing market. HEMS integrator services will act as specialists who implement the integration of network technology with residential services. We have also made the decision to accelerate and promote smart mansion (MEMS: Mansion (apartment) Energy Management System) implementation. The smallest MEMS units are HEMS, and we are also examining the application of ECHONET Lite in small-scale BEMS-like tenant buildings. Similar to the smart house market, this is also a market where we expect growth.

In summary, it is important to take quick action to jump-start the Smart House business and make international standardization possible by converting ECHONET Lite into a core open platform. At the same time, it is important to continue developing beneficial products and services for users (house dwellers). In order to make "gentle smart houses for home dwellers" a reality, it is important to produce "living innovations." Through linking IT equipment with home appliances and equipment, smart houses will provide a living environment that no one has ever experienced before. New living innovations have, as discussed, already started.

6.2 Overview of ECHONET Lite (International Open Standard for Home appliances and utilities) (by Murakami)

6.2.1 The purpose and aims of ECHONET Lite

Today, Japan not only faces difficulty in achieving its CO_2 emissions reduction goal established in the Kyoto Protocol of 1997 (6% reduction between 2008 and 2012 compared to the 1990 level) but it must significantly expand such reduction above and beyond 6% given further increases in CO_2 emissions.

In particular, emissions from the consumer sector, which includes office buildings and private households, rose 2.5 times compared to 1973 in FY2010. This makes the implementation of countermeasures in the household sector a matter of utmost importance. Moreover, the government has set the goals of reducing emissions by 25% compared to the 1990 level by 2020, and of bringing online photovoltaic facilities (which are highly effective in reducing CO_2 emissions) capable of producing approximately 28 million kW (approximately 20 times the current level) by 2020 (see Figure 6.6).

However, there are concerns that large-scale introduction of photovoltaic power generation will generate voltage build-up in transmission lines as power that is not consumed in households flows into the power distribution system, and that as a result the stable supply of high-quality power will be disrupted. Here, technical development is proceeding

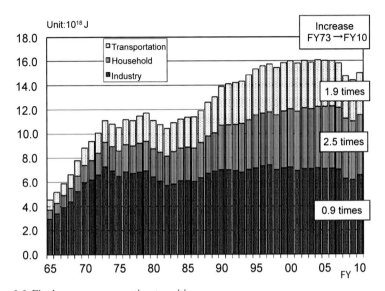

Figure 6.6 Final energy consumption transition.
Agency of Natural Resource and Energy: comprehensive energy statistics.

to address such concerns based on a new concept: the *smart grid*. On the household side, however, there is a need to guarantee the stable supply of power by maintaining a power supply-and-demand balance between the distribution system and households. Required here is a smart house concept that will further energy savings and ensure efficient use of generated photovoltaic energy by equipment installed in the house.

At the same time, as Japan's population continues to age, it is predicted that the percentage of households whose members are aged 65 years or older will reach 38% of all households by 2025. In fact, it is further predicted that the percentage of households whose members are at least 75 years old will reach 20%, meaning that one household in five will be an elderly household. This will further increase the importance of daily living assistance that allows senior citizens to manage their health and live with peace of mind (see Figure 6.7).

The ECHONET Consortium is improving standards and actively participating in national projects toward resolving these various issues, while also promoting activities to support the global development of home networks.

6.2.2 A smart house realized by ECHONET Lite

The first aim of ECHONET Lite is to develop and promote the adoption of a home network system using the electric appliances and equipment found in ordinary houses. ECHONET Lite is designed for use with application systems containing the same devices and functions found in ordinary houses, including single-family dwellings, duplexes, apartment buildings, dormitories, and condominiums for senior citizens. An example of a smart house is described in Figure 6.8.

ECHONET Lite also encompasses equipment systems for small office buildings and stores that are similar in terms of scale and system environment (cost, system

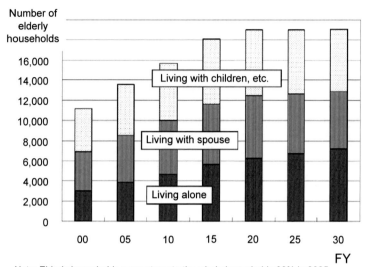

Note: Elderly households percentage to the whole households 38% in 2025.

Figure 6.7 Transition of the number of elderly households.
National Institute of Population and Social Security Research: future number of households in Japan (whole country estimation), March 2008.

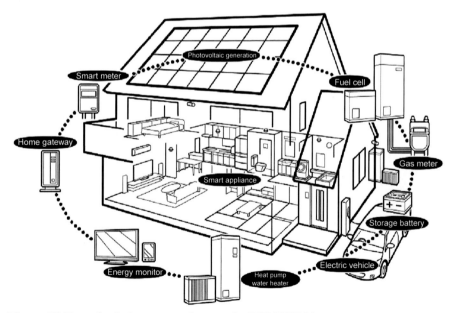

Figure 6.8 Example of a home network system by ECHONET Lite.

lifetime, functions, wiring restrictions, etc.) and that have yet to make substantial use of building or other facility management systems. Low-cost, easy-to-use subnetwork systems can be built regardless of building size: entire-building systems for small buildings, or floor-by-floor systems in larger structures.

The ECHONET logo represents people as the central player of ECHONET surrounded by systems and the environment. The continuous line reflects how human life is inextricably linked with systems and the environment. Blue represents the color of the sea which nurtures life, the color of a clear sky spreading toward the future, and the color of a clean environment which is the target of ECHONET. The Logo will be marked on home appliances which meet the ECHONET specification.

ECHONET

ECHONET™ エコーネット™ ECHONETLite™ ECHONETReady™

and the above logos are the registered trademarks of the ECHONET Consortium respectively.

Figure 6.9 Logo of ECHONET Consortium.

6.2.3 Overview of ECHONET Lite

The main features of ECHONET Lite, which can create a smart house, are the following:

- Selects a communication media by setting environment, function requirements, development environment, etc. ECHONET Lite makes best use of the technology of communication media that the user wants to adopt.
- Allows use of detailed rules (ECHONET device objects) for all types of household equipment (92 categories) – including household devices, residential appliances, sensors, and health management devices – as ECHONET devices.
- Allows easy installation and changing of systems through plug-and-play functions.
- Permits free combination of transmission media (e.g., wireless media, wired media, etc.) according to the installation environment.
- Applicable for ordinary households, multidwelling buildings, shops, and small- and medium-sized office buildings.

ECHONET Lite is an "open standard." All specifications are open on the Web (http://www.echonet.gr.jp/english/spec/index.htm). All control commands related to energy management service have been already standardized as IEC 62394. These international standard documents are open on the Web (http://webstore.iec.ch/webstore/webstore.nsf/artnum/048660!opendocument). Logo is as Figure 6.9.

6.2.4 Explanation of ECHONET Lite

The ECHONET Lite protocol is transport free so that the system integrator can select communication media they want to use, making it easy for ECHONET Lite to construct a home network system.

ECHONET Lite mainly specifies the communication middleware (message structure, communication sequence) and control commands of all types of household equipment.

1. ECHONET Lite communication middleware

The ECHONET Lite Communication middleware is responsible for the following:

- Processing the communication protocols needed to facilitate processing when application software is remotely controlling or monitoring a device in an equipment system

• Storing the data needed to process communication protocols
• Managing data such as device statuses

ECHONET Lite specifies this communication protocol. Of the data stored by this middleware, the data and access procedures that are disclosed to other devices are expressed as device objects and specified as ECHONET Lite object definitions.

2. ECHONET control commands

ECHONET Lite also specifies many kinds of control commands, which are called "device objects." Some control commands have been an international standard specified as IEC 62394. The device object is a logical model of the information held by equipment devices or home electrical appliances such as sensors, air conditioners, and refrigerators, or of control items that can be remotely controlled. The interface form for remote control is standardized. Since this device object is specified for each type of device, even products of different manufacturers can be remote controlled in exactly the same way if they are of the same device type. More specifically, the information and control target of each device is specified as a property, and the operating method (setting and browsing) is specified as a service.

The example of devices whose control commands have been specified is described in Table 6.1.

Table 6.1 Equipment stipulated in the ECHONET device objects

Class group code	Class group	Examples
0x00	Sensor-related device class group	Human detection sensor, electric energy sensor, electric energy sensor, etc.
0x01	Air conditioner-related device class group	Home air conditioner, ventilation fan, air cleaner, humidifier, package-type commercial air conditioner, etc.
0x02	Housing/facilities-related device class group	Electrically operated shade, bidet-equipped toilet (with electrically warmed seat), home solar power generation, floor heater, fuel cell, storage battery, electric energy meter, distribution board metering, smart electric energy meter, smart gas meter, general lighting, etc.
0x03	Cooking/household-related device class group	Refrigerator, combination microwave oven (electronic oven), cooking heater, rice cooker, washer and dryer, etc.
0x04	Health-related device class group	Weighing machine
0x05	Management/ operation-related device class group	Switch (JEM-A/HA)
0x06	Audiovisual-related device class group	Display, television

6.3 Overview of Smart House Businesses 1: Panasonic activities (by Murakami)

6.3.1 Panasonic to develop a full-scale smart home energy management system business compliant with ECHONET Lite

Panasonic Corporation began the full-scale development of its Smart Home Energy Management System (SMARTHEMS®[1]) business compliant with ECHONET Lite on October 21, 2012, in Japan. It has also launched Artificial Intelligence Smart Energy Gateway (AiSEG®[2]), which is a core component of SMARTHEMS®, connecting the electrical equipment and home appliances in the system. Panasonic SMARTHEMS® will evolve to provide a comfortable lifestyle with energy creation, storage, and automatic energy saving (Figure 6.10).

Being equipped with ECHONET Lite, SMARTHEMS® not only visualizes the consumption of energy in the household but can also control different types of home electrical equipment through the core component, AiSEG®. Panasonic will gradually increase its lineup of products that can be linked through HEMS, including air conditioners, IH cooking equipment, and EcoCute.[3]

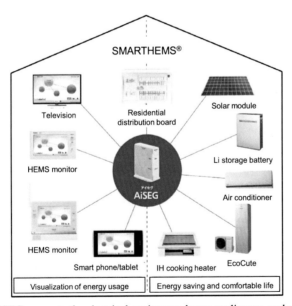

Figure 6.10 AiSEG connects the electrical equipment, home appliances, and some kinds of monitors in SMARTHEMS®.

---

[1] SMARTHEMS is a trademark of the Panasonic Group.
[2] AiSEG is a trademark of the Panasonic Group.
[3] A general industry term used by electric power companies and manufacturers for heat pump hot water supply systems that use natural refrigerants.

SMARTHEMS® has the following features:

1. Visualizes data of solar power generation systems, storage batteries, electricity, gas, and water supplies.
2. Automatically controls the operation of compatible equipment with ECHONET Lite, contributing to energy saving and a more comfortable life.
3. Cloud computing service ensures compatibility with expanding product lines and offers adaptability to changes in the social landscape (Figure 6.10).

In order for SMARTHEMS® to respond to future developments, such as changes in the electricity rating system, and enable the long-term use of the system, firmware updates will be provided for the system via cloud computing service.

Panasonic will continue to improve and expand its HEMS, SMARTHEMS®, to create, store and use energy efficiently, and manage it smartly by linking home electrical equipment and home appliances.

6.4 Overview of Smart House Businesses 2: Toshiba activities (by Hirahara and Minemura)

6.4.1 Popularizing HEMS world with ECHONET Lite™ using FEMINITY-Club server

More and more energy is consumed these days, especially in households; a trend that increases the need for swiftly spreading countermeasures such as HEMS. HEMS allows M2M home networking through communication and control protocols, some of which are standardized, such as the IEC standard called ECHONET Lite™. ECHONET Lite™, which has also been adopted by Toshiba, enables us to apply various services to home devices.

Those examples are "Visualization" of energy consumption, "Energy saving advice" based on use history data, and "Store electricity" or "Increase capacity" to or from a prepared storage battery, responding to varying electricity consumption in daily life. The controlling is not limited to those equipments, but it is possible to expand to various devices in each room, and even a comprehensive linked control of devices including air conditioning and lighting. Such prebundled controlling can simply be "Going out" or "Going to bed" that fits with each life scene. Moreover, users can remotely monitor and control devices such as "Air conditioning," "Lighting," "Floor heating," and "Water heating," as well as "Visitor noticing" and "Caution messaging."

Since 2002, Toshiba has been releasing IT consumer electronics, called ECHONET™-ready devices such as air conditioners, refrigerators, and washing machines. These evolvable devices have been designed to be easily equipped with HEMS capabilities, when users additionally purchase ECHONET™ middleware adaptors for connection.

Currently, the HEMS market has been brisk owing to spreading acknowledgment of value-adding HEMS for advancing living styles in residential complexes and detached housings, with the backing of a MEMS subsidy system that has been promoted by the Ministry of Economy, Trade and Industry. There are already many practical market examples, including residential complexes in Japan as well as overseas.

To realize interoperable HEMS among various home appliances that have been made by numerous companies, the designated Website provides the command

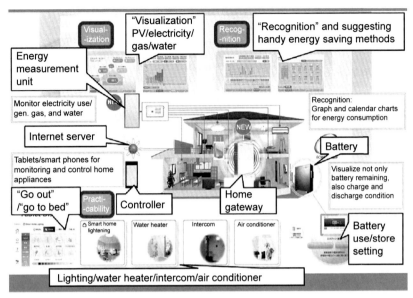

Figure 6.11 Outline of FEMINITY Services using the home network system.

specifications for monitoring and controlling in underlying ECHONET Lite™. Categories of applicable devices have already surpassed 100 in number. To support device interoperability among hundreds of makers, the HEMS Certification Support Center has provided test environments with communication counterparts available in the market. To simplify and shorten implementation of ECHONET Lite™ to target devices, various SDK kits are freely available (Figure 6.11).

The system provides "Visualization" of electricity, gas, and water consumption, as well as PV electricity generation with a storage battery condition. Furthermore, the system does "Recognition" by displaying graphs and schedulers, and "Activation" by activating HEMS control through display devices such as tablets and smartphones (Figure 6.12).

The system reduces peak electricity demands through automated control of large-capacity storage batteries, air conditioning, and other facilities via HEMS for each unit and MEMS. Applications are not limited to "Visualization" of energy but have been expanded to "Recognition." Park Tower Nishishinjuku Emsport (http://www.mitsuifudosan. co.jp/corporate/news/2012/0919/, http://www.mitsuifudosan.co.jp/english/corporate/ news/2012/0712/index.html).

*ECHONET and ECHONET Lite are trademarks of the ECHONET Consortium.

6.5   Open source software and related activities for ECHONET Lite with web technology (by Owada)

Now we introduce two open source software products that provide easier access to the ECHONET Lite network. One is OpenECHO, a Java class library that directly implements ECHONET Lite device objects. The other is KadecotCore, an Android home

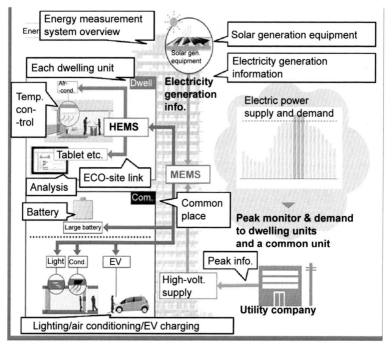

Figure 6.12 Home energy management system introductory example "Park Tower Nishi-Shinjuku Emsport" (total number of units: 179/Shinjuku Ward (*)/Tokyo, Japan) *The metropolitan government office is in Shinjuku Ward.

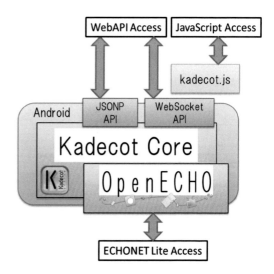

Figure 6.13 System architecture.

server application that provides WebAPIs to control home networks from HTML5 applications. We also briefly describe how to use these APIs. We conclude by stating that our activities are key elements for promoting and cultivating the field of home network applications (see Figure 6.13).

Figure 6.14 Screenshot of OpenECHO for processing.

6.5.1 *OpenECHO: a Java class library for ECHONET Lite*

OpenECHO (OpenECHO, https://github.com/SonyCSL/OpenECHO) is a low-level Java class library that directly implements the IPv4-based ECHONET Lite protocol. Since it is written in pure Java, it runs on various platforms such as Windows, Mac, Linux, and Android. The user of this library should be able to use the Java language and link this library to their own software project. This library is useful, for example, for the development of custom-made home appliances or an emulator of ECHONET Lite.

For introductory users to experience this library, there is also a specially compiled version for Processing (OpenECHO for Processing, http://smarthouse-center.org/sdk/detail/2). (Processing is a popular Java programming environment developed at the Massachusetts Institute of Technology (MIT) (Processing, http://processing.org).) The Processing version is fully documented with sample codes in Japanese and English. This was developed and released with the support of the HEMS Certification Center at the Kanagawa Institute of Technology (see Figure 6.14).

6.5.2 *KadecotCore: an Android WebAPI server for ECHONET Lite*

Recently, a significant portion of consumer applications have been written in HTML5, which runs on all major Web browsers. Although there are some drawbacks with HTML5, such as relatively low performance or source code visibility, HTML5 achieves high portability and suitability for mixing with other Web services (called mash-up). Therefore, there is almost no doubt that HTML will be even more commonly used in future. On the other hand, the ECHONET Lite network cannot be directly accessed from HTML5 since it is usually implemented on a lower-level IP protocol.

KadecotCore is intended to bridge this gap. On the one hand, it acts as a Web server that provides HTML-level access for devices. On the other hand, it communicates with ECHONET Lite devices in the lower level using the OpenECHO library.

For the HTML level, KadecotCore provides two kinds of WebAPI: JSONP API and WebSocket API. WebSocket API is further wrapped by a JavaScript library called kadecot.js to provide easy access for JavaScript programmers. JSONP API access is very simple and easy, but the connection is established and disconnected every time access occurs. Therefore, the change in device state cannot be detected in real time (polling is the only solution to monitor state changes). On the other hand, WebSocket API and its wrapping JavaScript library handle a constant connection to monitor and handle real-time changes in the device network. The source code is available on GitHub (KadecotCore, https://github.com/SonyCSL/KadecotCore).

Although KadecotCore is an Android application, it has not been distributed from Google Play (Google Play, https://play.google.com/store) yet. Instead, we distribute a larger application, called Kadecot (Kadecot, http://kadecot.net/), from Google Play. Kadecot links KadecotCore as a library. Since KadecotCore is configured as a library project, it can be linked from another project. The additional functions of Kadecot over KadecotCore include support for Sony Blu-ray recorders, commercial infrared controllers, and connection to online services (Kadecot online page, http://apps.kadecot.net/). ECHONET Lite-related APIs are the same for Kadecot and KadecotCore. Therefore, simply downloading Kadecot from Google Play is a convenient solution for those who are not interested in source codes.

Figure 6.15 shows the main panel of KadecotCore/Kadecot.

In the main panel, the physically available or software-emulated devices are listed as icons. Simple operations such as turning on/off devices can be performed by simply pressing the corresponding icon. The behavior depends on the device type. For example, a light or an air conditioner turns on/off, an electronic curtain opens or closes, while a temperature or humidity sensor refreshes their values. The user can assign a unique nickname to each device. The nicknames are also used by APIs for identification purposes.

To turn on API server functionality, press the "Settings" button located on the top-right to open the settings panel and select the "Run JSONP Server" or "Run WebSocket Server" checkbox.

1. JSONP API

JSONP is an HTTP-based common form of the WebAPI interface. The convenience of the JSONP interface lies in its ability to access resources on different domains (called cross-domain access), as well as its seamless integration with the JavaScript language. To access the KadecotCore JSONP server, the client should know the server IP address displayed either in the "Settings" panel or in Android's notification area. KadecotCore JSONP API uses port 31413.

Figure 6.15 KadecotCore/Kadecot's main panel.

Once the server IP address is known, accessing the server is easy. For example, the list of recognized devices can be obtained by accessing the following URL:

```
http://%SERVER IP%:31413/call.json?method=list
```

where %SERVER IP% is the IP address of the server. The response should look like this:

```
{"result":[
  {"active":[[se should be like this Litee":deviceName"eviceNamese ",
  "nickname":"Controller", "option":{},"deviceType":"0x05ff"},
```

```
{{05ffTypese should be like
 this:followingdevicename"evPowerDistributionBoardMetering "o,
 "nickname":"PowerDistributionBoardMetering",
 "option":{},"deviceType":"0x0287"},
 {{0x0287peeTypeonBoardMeteringis:followingdevicename
 "evHomeAirConditioner",
 "nicknameondHomeAirConditioner", "optionrCo },
 "deviceType": "0x0130"}
 ]}
```

In the above example, three devices are recognized: a controller device, which is KadecotCore itself, a power panel, and an air conditioner.

Obtaining current device properties is also easy. For example, the following URL requests current power status:

```
http://%SERVER IP%:31413/call.json?method=get&params=
[HomeAirConditioner,0x80]
```

Note the "method" parameter is set as "get," while the "params" parameter is an array of the device nickname and the epc code (0x80 is the code for power status property. See the appendix of the ECHONET Lite specification).

Changing the device status is also easy. The following code turns on the power of the air conditioner:

```
http://%SERVER IP%:31413/call.json?method=set&params=
[HomeAirConditioner,[0x80,[0x30]]]
```

0x30 is the code for turning on the power. The successful response looks like this:

```
{"result":{"nickname": "HomeAirConditioner", "property":[{ "value":
[48], "success":true, "name": "0x80"}]}}
```

Value is the set value (48 in hexadecimal format is 0x30.)

As shown above, the JSONP API is very simple and easy to use. On the other hand, this interface is open to all applications in the same network. If there is a harmful application, the life log may be stolen by the software or the appliances may be manipulated in a malicious and dangerous way. Please be reminded that JSONP server functionality is only for evaluation purposes. WebSocket API is preferred for actual use.

2. WebSocket API

The WebSocket client is implemented in major Web browsers that support HTML5. It is a standard TCP/IP socket with a slightly complicated handshaking phase during the initial connection. KadecotCore supports WebSocket server functionality to achieve constant connection to client Web browsers. The basic structure is the same as a JSONP server. The port number is 41314.

The communication protocol over WebSocket is based on a simple standard called "JSON-RPC" (JSON-RPC, http://json-rpc.org/). This format is for achieving a remote function call by exchanging a JSON object. JSON-RPC requires a JSON object to implement at least two members: "method" (for function name) and "params" (for function parameters). We added "version" and some context-dependent members to retain future compatibility and solve other implementation issues. The WebSocket API consists of server management methods and device management methods. Due to a lack of space, a detailed description of the API set is beyond the scope of our focus.

Since WebSocket API is a mixture of server functions and device functions, KadecotCore is also equipped with a JavaScript wrapping library called kadecot.js. With kadecot.js, a programmer can concentrate only on device accesses. kadecot.js also provides a unique device access methodology. First, the client application declares which devices and properties it is interested in. This information is called the "manifest." The manifest is given to kadecot.js, and eventually to the KadecotCore server. Then the system gives back an "access object" to the client application based on the manifest. The access object hierarchically contains special member variables for accessing the real device. For example, if the programmer substitutes a value to the access object member, it is recognized as setting a new value to a device property. The corresponding device should work or return the answer value. Similarly, referring to the value gives the current state of the device.

6.5.3   Conclusions and future work

We briefly described two ECHONET Lite-related open source software products: OpenECHO and KadecotCore. We also described two kinds of WebAPIs with one wrapping library, provided by KadecotCore. These software products are freely available on the Web from source codes.

Index

Note: Page numbers followed by "f" and "t" indicate figures and tables respectively.

Printed in the United States
By Bookmasters